HISTORY OF FUNCTIONAL ANALYSIS

NORTH-HOLLAND
MATHEMATICS STUDIES   **49**

Notas de Matemática (77)

Editor: Leopoldo Nachbin

Universidade Federal do Rio de Janeiro
and University of Rochester

History of Functional Analysis

JEAN DIEUDONNÉ

Professeur honoraire à la Faculté des Sciences
de Nice, France

NORTH-HOLLAND — AMSTERDAM · NEW YORK · OXFORD

ISBN: 0 444 86148 3

First edition 1981
Second printing 1983

Published by:
ELSEVIER SCIENCE PUBLISHERS B.V.
P.O. Box 1991
1000 BZ Amsterdam
The Netherlands

Sole distributors for the U.S.A. and Canada:
ELSEVIER SCIENCE PUBLISHING COMPANY, INC.
52 Vanderbilt Avenue
New York, N.Y. 10017
U.S.A.

Library of Congress Cataloging in Publication Data

Dieudonné, Jean Alexandre, 1906-
    History of functional analysis.

    (Notas de mathemática ; 77) (North-Holland
mathematics studies ; 49)
    Bibliography: p.
    Includes indexes.
    1. Functional analysis--History. I. Title.
II. Series.
QA1.N86 no. 77 [QA320]   510s   80-28960
ISBN 0-444-86148-3 (Elsevier) [515.7'09]

Transferred to digital printing 2005

TABLE OF CONTENTS

INTRODUCTION

One may give many definitions of "Functional Analysis". Its
name might suggest that it contains all parts of mathematics
which deal with functions, but that would practically mean <u>all</u>
mathematical Analysis. We shall adopt a narrower definition:
for us, it will be the study of topological vector spaces and
of mappings $u: \Omega \to F$ from a part Ω of a topological vector
space E into a topological vector space F, these mappings
being assumed to satisfy various algebraic and topological conditions. A moment of reflection shows that this already covers
a large part of modern Analysis, in particular the theory of
partial differential equations.

Functional Analysis thus appears as a rather complex blend of
Algebra and Topology, and it should therefore surprise no one
that the development of these two branches of mathematics had
a strong influence on its own evolution. As a matter of fact,
it is almost impossible to dissociate the early history of
General Topology (and even of the set-theoretic language) from
the beginnings of Functional Analysis, since the sets and spaces
which (after the subsets of $\mathbb{R}^n$) attracted most attention consisted of <u>functions</u>.

With regard to Algebra, as the most frequently studied mappings
between topological vector spaces are <u>linear</u>, it is quite natural that linear Algebra should have greatly influenced
Functional Analysis. In fact, at the end of the XIXth century,
the old idea that infinitesimal Calculus was derived from the

1

algebraic "Calculus of differences" by a "limit process" began
to acquire a more precise and more influential form when Volterra applied a similar idea to an integral equation

$$(1) \qquad \int_a^y \varphi(x)H(x,y)dx = f(y)$$

for an unknown function φ, the functions f and H being
continuous in $[a,b]$ and $[a,b] \times [a,b]$ respectively, with
$f(a) = 0$. He divides $[a,b]$ into n subintervals by the
points $y_k = a+k\,\frac{b-a}{n}$ $(1 \leq k \leq n)$, replaces y in (1) by
these n values, and the integral by the corresponding Riemann
sums, which gives him a system of n linear equations

$$(2) \qquad \begin{cases} h_{11}z_1 & = b_1 \\ h_{21}z_1 + h_{22}z_2 & = b_2 \\ \cdots\cdots\cdots\cdots\cdots\cdots\cdots\cdots\cdots \\ h_{n1}z_1 + h_{n2}z_2 + \cdots + h_{nn}z_n = b_n \end{cases}$$

with $h_{jk} = H(y_j,y_k)$, $z_k = \varphi(y_k)$ and $b_k = f(y_k)$; the integral equation (1) was thus considered as obtained from systems
(2) by a limit process when the number of unknowns became infinite.

Unfortunately, linear Algebra, as it was understood in the
XIXth century (and even much later) did not readily lend itself to affording a good guidance to such generalizations.
Its own evolution had been very slow and painful, stretching
over 130 years, and in a succession of stages which, to our
eyes, is exactly the reverse of the logical sequence of notions, namely

linear equations

↓

determinants

↓

linear and bilinear forms

↓

matrices

↓

vector spaces and linear maps

In spite of the unsuccessful efforts of Grassmann and Peano, the intrinsic aspects and the geometric point of view in linear Algebra remained in the background until 1900; one would readily speak with Cayley (1843) of vectors and linear subspaces, but they were invariably considered as parts of some $\mathbb{R}^n$; in other words, everything in a vector space was always referred to a <u>fixed basis</u>, and linear maps were only handled through their <u>matrices</u> corresponding to these bases. The various "reduction" theorems were known in 1880, but only through complicated computations of determinants, and without any geometric interpretation. Furthermore, Frobenius, who had been the most influential mathematician in building up a synthesis of the linear Algebra of his time, had unfortunately taken a step backward (even with respect to Cayley) by electing to work systematically with bilinear forms $\sum_{p,q} a_{pq} x_p y_q$ instead of working with matrices (a_{pq}). Finally, before 1930 nobody had a correct conception of duality between finite dimensional <u>vector spaces</u>; even in van der Waerden's book (1931), such a vector space and its dual are still <u>identified</u>.

All this was to weigh heavily on the evolution of linear Functional Analysis; in particular it followed (over a shorter span of years) the same unfortunate succession of stages through

which linear Algebra had to go; and it is only after it was
realized that the current conception of vectors as "n-tuples"
could not possibly be extended to infinite dimensional function spaces, that this conception was finally abandoned and
that genuinely geometrical notions won the day.

The diagram at the end of this Introduction tries to depict
graphically in some detail the successive stages of the history of Functional Analysis, by mentioning the actions and re-
actions of the various parts of mathematics which took part in
it. If one were to reduce this complicated history to a few
key words, I think the emphasis should fall on the evolution
of two concepts: spectral theory and duality. Both of course
stem from the very concrete problems encountered in the solution of linear equations (or systems of linear equations),
where the unknowns are functions. The basic concepts of spectral theory: eigenvalues, eigenfunctions and expansions in
series of such functions were already known at the beginning
of the XIXth century, in the theory of Fourier series; they
would form the model on which all further advances were patterned. But it took more than 60 years of strenuous efforts
to extend the theory from the Sturm-Liouville problem in ordinary differential equations to the partial differential equation of the vibrating membrane. It was gradually realized
that the heart of the matter lay, not in the differential (or
partial differential) equations themselves, but in integral
equations associated to them; at first they were not explicitly written down, so that one can only speak of "crypto-integral" eqations, to designate the use of methods resting on

evaluations of integrals, and which only later emerged as

standard methods in the theory of integral equations.

The remarkable feature of this history is that, after such a

slow incubation period, so to speak, spectral theory, in the

span of a few years, reached complete maturity, giving birth

in the process to the concept of linear duality, which began

at last to be understood by analysts, before becoming later

familiar to all mathematicians by a kind of backlash effect.

What is interesting in this rapid advance is that it was ac-

complished in a series of what one may call discrete jumps,

in each of which the decisive step was to ignore the special

features of the problem under consideration, and to make it

accessible by inserting it into a more general context.

The first of these "discontinuities" occurred in 1896-1900,

when Le Roux, Volterra and Fredholm, instead of working on

the special integral equations studied by their predecessors

(Abel, Liouville, Beer-Neumann), elected to use minimal as-

sumptions on the kernels, and in so doing discovered that the

theory was far simpler than it was generally thought.

The second step was taken by Hilbert in his 1906 paper, sub-

ordinating the too special theory of symmetric integral equa-

tions to the much more general concept of infinite "bounded"

quadratic forms, which turned out to provide the frame needed

for all subsequent progress in ordinary and partial differen-

tial equations.

The contemporary discovery of the Lebesgue integral, and the

geometric and topological concepts introduced by Fréchet in

Analysis immediately led Hilbert's successors to translate his

results into the language of what we now call Hilbert space,
linking euclidean geometry to integration theory, and making
possible the discussion of the most general system of linear
equations in such a space.

This in turn led F. Riesz in 1910-1913 to introduce L^p and
ℓ^p spaces for any exponent p such that $1 < p < +\infty$, and
to discover the natural duality between the <u>different</u> spaces
L^p and L^q with $\frac{1}{p} + \frac{1}{q} = 1$, in sharp distinction from the
muddleheaded ideas on the matter, which the accidental self-
duality of Hilbert space had failed to dispel.

But although F. Riesz, in the treatment of systems of linear
equations in ℓ^p spaces, was the first to obtain a condition
which later was seen to consist in a particular application
of the Hahn-Banach theorem, he failed to visualize that condition as amounting to an extension property of a continuous
linear form defined on a subspace. This fourth "jump" was only
accomplished by Helly in 1921, again by generalizing the theory of systems of linear equations from the special ℓ^p spaces
to <u>any</u> normed subspace of $\mathbb{C}^N$. After that, only two more steps
were needed to reach the present status of the theory, with
the passage to general normed spaces (together with the use
of transfinite induction) by Hahn and Banach and a little
later the extension of duality theory to locally convex spaces
during the period 1935-1945.

This process of successive generalizations may thus have reached a point of diminishing returns around the middle of the
century. Inasmuch as we are able to judge from events probably too recent to allow a proper perspective, the theory of

topological vector spaces, after 1950, has stabilized as one of the standard tools of modern mathematics, together with linear and multilinear Algebra, General Topology and measure theory. The advances which have been achieved during the last 30 years mainly consist in new imaginative ways to use the fundamental tools of Functional Analysis, either in theories where they had not been applied before, such as differential geometry and differential topology (K-theory, theory of the Atiyah-Singer index, foliations), or in the construction of more powerful methods to handle functional equations (distributions, Sobolev spaces, pseudo-differential operators and their generalizations).

This volume grew out of a series of a lectures which I gave in Rio de Janeiro in 1979, at the invitation of Prof. Jorge Alberto Barroso of the Universidade Federal do Rio de Janeiro, to whom go my most heartfelt thanks. I am also very grateful to him for the pains he took in supervising the preparation of the manuscript for publication.

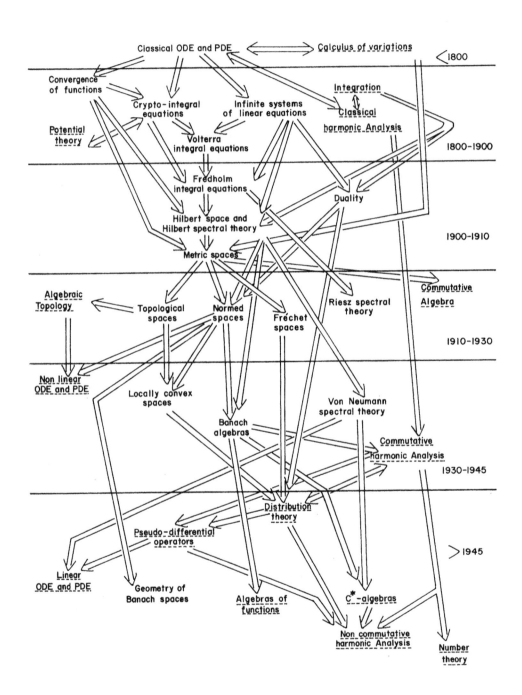

CHAPTER I

LINEAR DIFFERENTIAL EQUATIONS AND THE STURM-LIOUVILLE PROBLEM

§1. Differential equations and partial differential equations in the XVIIIth century.

Until around 1750, the notion of _function_ of one variable was a very hazy one. The domain where it was defined was very seldom described with precision; it was tacitly assumed that around each point x_o, the function was equal to a power series in $x-x_o$ and its derivatives were obtained by taking the derivatives of each term of the series. To solve a differential equation of order n

(1) $$y^{(n)} = F(x,y,y',y'',\ldots,y^{(n-1)})$$

one would therefore substitute in (1) for y and its derivatives a power series $\sum_{k=0}^{\infty} c_k(x-x_o)^k$ and its derivatives, and identify the series on both sides, which would determine each c_k for $k \geq n$ as a function of $c_o, c_1, \ldots, c_{k-1}$; the solution thus depended on n arbitrary parameters $c_o, \ldots, c_{n-1}$. The very few cases in which it was possible to write explicitly the solution by means of primitives of known functions (such as the linear equation $y' = a(x)y + b(x)$ of order 1) were already known at the end of the XVIIth century. After 1760 began the first general study of linear equations of arbitrary order

9

(2) $L(y) \equiv y^{(n)} + a_1(x)y^{(n-1)} + \ldots + a_n(x)y = b(x).$

D'Alembert observed that the knowledge of a particular solution of the equation and of all solutions of the homogeneous
equation $L(y) = 0$ yields by addition all solutions of (2).
A little later, Lagrange [135, vol. I, p.474] showed that the
general solution of $L(y) = 0$ may be written $\sum_{k=1}^{n} c_k y_k$ where
the c_k are arbitrary constants, and the y_k ($1 \le k \le n$) particular solutions (which he tacitly assumed to be linearly
independent). Then, by his famous method of "variation of
constants" [135, vol. IV, p.159], he showed how to obtain
also the solutions of (2) when the y_k were known: the solution is written in the form $y = \sum_{k=1}^{n} z_k y_k$, where the z_k
are unknown functions, subject to $n-1$ linear relations

(3) $\sum_{k=1}^{n} z'_k y_k^{(\nu)} = 0$ $(0 \le \nu \le n-2).$

These conditions imply that $y^{(\nu)} = \sum_{k=1}^{n} z_k y_k^{(\nu)}$ for $0 \le \nu \le n-1$;
replacing y by $\sum_{k=1}^{n} z_k y_k$ in (2) and using the fact that
the y_k satisfy $L(y_k) = 0$, one obtains for the z'_k another
linear equation

(4) $\sum_{k=1}^{n} z'_k y_k^{(n-1)} = b(x)$

from which, by the Cramer formulas, one can compute the z'_k
($1 \le k \le n$) and the problem is thus reduced to computing their
primitives.

Lagrange also introduced [135, vol. I, p.471] the notion of
<u>adjoint</u> of a linear differential operator L, which was to

acquire great importance later: he showed that there exists
a linear differential operator M satisfying an identity

(5) $$z \, L(y) - y \, M(z) = \frac{d}{dx} \left(B(y,z) \right)$$

where B is bilinear in $(y, y', \ldots, y^{(n-1)})$ and $(z, z', \ldots, z^{(n-1)})$,
constituting a generalization of the classical "integration
by parts"; he deduced from that formula that if a solution z
of $M(z) = 0$ was known, solutions of $L(y) = 0$ could be
obtained by solving an equation $B(y,z) = $ Const. of order n-1.

Partial differential equations were not considered until the
middle of the XVIIIth century, in connection with problems of
Mechanics or Physics and then they were of order 2 at least
(see §2). The study of partial differential equations of first
order was only begun by Euler and Lagrange after 1770. Euler
was able to solve a few particular equations, and then Lagrange
found general methods which enabled his followers, Charpit and
Monge, to reduce the solution of a general equation of first
order

(6) $$F\left(x, y, z, \frac{\partial z}{\partial x}, \frac{\partial z}{\partial y}\right) = 0$$

to the solution of a system of ordinary differential equations,
an idea which was developed later by Cauchy in his concept of
"characteristic curves".

§2. Fourier expansions.

In 1747, d'Alembert gave the first mathematical treatment of
the general problem of the small vibrations of a string of
length a, fixed at each extremity; the string moves in a plane

where the axis Ox is along the position of the string at
rest, the segment $0 \leq x \leq a$; if $y = u(x,t)$ is the equation
of the string at time t, d'Alembert shows that, if $u(x,t)$
remains small, it satisfies the equation

$$(7) \qquad \frac{\partial^2 u}{\partial t^2} = c^2 \cdot \frac{\partial^2 u}{\partial x^2}$$

where c is a known function of x alone, and is constant
if the density of the string is constant. When c is constant, taking $X = x-ct$ and $Y = x+ct$ as new variables re-
duces the equation to $\frac{\partial^2 u}{\partial X \partial Y} = 0$, and d'Alembert concluded
that the solution of (7) is given by

$$(8) \qquad u(x,t) = f(x-ct) + g(x+ct)$$

where f and g are "arbitrary" functions. A year later,
Euler interpreted this result as meaning that (for $c=1$)
$u(x,t)$ was known once the two functions of x,

$$(9) \qquad u(x,0) = \varphi(x), \qquad \frac{\partial u}{\partial t}(x,0) = \psi(x)$$

were prescribed, the value of $u(x,t)$ being explicitly
given by

$$(10) \qquad u(x,t) = \frac{1}{2}(\varphi(x-t) + \varphi(x+t)) + \frac{1}{2}\int_{-t}^{t} \psi(x-\xi)d\xi$$

(Euler only gives a geometric construction equivalent to this
formula). Now it was well known experimentally that $\varphi(x)$
could be quite different from an analytic function, for instance it could have no derivative at some points, and this
led Euler to introduce, in addition to what he called "continuous" functions (i.e. analytic functions in our sense) more

general ones which he baptized "mechanical" without giving
their precise definition (from the context they seem to be
piecewise twice differentiable functions in our terminology).
On the other hand, already in 1715, B. Taylor, by a direct
argument which did not use equation (7), had concluded that
(when c is constant) for any integer $n \geq 1$, the functions

$$(11) \qquad u_n(x,t) = \sin \frac{n\pi x}{a} \cos \frac{n\pi ct}{a}$$

represented vibrations of the string, namely for $n = 1$ the
"fundamental" tone, and for $n = 2,3,\ldots$, its "harmonics".
As it was well known that the sound emitted by a vibrating
string was in general a mixture of several "harmonics", Daniel
Bernoulli, in 1750, proposed that the general solution (10)
could also be written as a series

$$(12) \qquad u(x,t) = \sum_{n=1}^{\infty} a_n \sin \frac{n\pi x}{a} \cos \frac{n\pi c}{a} (t-\beta_n)$$

for suitable values of the a_n and β_n. However, in 1753,
Euler observed that this would imply that an arbitrary "mechanical" function defined in an interval $-a \leq x \leq a$ could
be written as a series

$$(13) \quad \frac{a_o}{2} + a_1 \cos \frac{\pi x}{a} + b_1 \sin \frac{\pi x}{a} + a_2 \cos \frac{2\pi x}{a} + b_2 \sin \frac{2\pi x}{a} + \ldots$$

and he believed that such a series of analytic functions could
only represent an analytic function. His opinion was shared
(with some variations) by almost all other mathematicians of
his time, and no progress was made on this question until the
beginning of Fourier's work on the theory of heat (see [65,
(2), t. XI_2, pp.273-300]). Having to solve equations such as

$$(14) \qquad \frac{\partial^2 u}{\partial x^2} + \frac{\partial^2 u}{\partial y^2} = 0$$

$$(15) \qquad \frac{\partial^2 u}{\partial x^2} - \frac{\partial u}{\partial y} = 0$$

for various boundary conditions, he systematically looks for solutions of the form $u(x,y) = v(x)w(y)$ and, following D. Bernoulli, wants to obtain the most general solution as series whose terms are these particular ones. In so doing, he is brought back to the problem of expressing a function f as a series (13), but this time he adds to D. Bernoulli's argument the formulas giving actually the values of the coefficients a_n, b_n

$$(16) \quad a_n = \frac{1}{\pi} \int_{-\pi}^{\pi} f(x) \cos nx \, dx, \qquad b_n = \frac{1}{\pi} \int_{-\pi}^{\pi} f(x) \sin nx \, dx$$

(when $a = \pi$) which as a matter of fact had already been obtained by Clairaut and Euler, without realizing their interest. Using these formulas Fourier was able to show on many examples of non analytic functions that the corresponding Fourier series converged to $\frac{1}{2}(f(x+) + f(x-))$, and expressed his conviction that this was true for "arbitrary" functions, although his attempts and those of Cauchy to prove that result were unsuccessful and the first proof for a piecewise monotonic and piecewise continuous function was only given by Dirichlet in 1829. One should also mention in that connection that in 1799, Parseval had given the formula

$$(17) \qquad \frac{a_o^2}{2} + \sum_{n=1}^{\infty} (a_n^2 + b_n^2) = \frac{1}{\pi} \int_{-\pi}^{\pi} (f(t))^2 \, dt$$

by a purely formal computation, without any proof of convergence.

These results gave the impetus to the vast theory of __trigonometric series__, which was to be one of the main concerns of most analysts in the XIXth century, centered around the criteria of convergence of such series and the relations between its sum and its coefficients. The evolution of that theory was closely linked to a gradual precision and deepening of the notions of set of real numbers, of function and of integral. But before 1920 there was not much contact between that theory and the development of Functional Analysis as we understand it.

On the contrary, other results of Fourier in his __Theory of heat__ triggered the birth of __spectral theory__. For instance [67, vol. I, p.304] he shows that the "cooling off" problem for a solid sphere of radius r, when one assumes spherical symmetry for the problem, is governed by the partial differential equation

(18)
$$\frac{\partial u}{\partial t} = k\left(\frac{\partial^2 u}{\partial x^2} + \frac{2}{x}\frac{\partial u}{\partial x}\right)$$

with the "boundary conditions" that $u(x,t)$ must remain finite when x tends to 0, and satisfy the relation

(19)
$$\frac{\partial u}{\partial x} + hu = 0 \quad \text{for} \quad x = r \quad \text{and all} \quad t,$$

where h and k are constants. Using his favorite method of "separation of variables", Fourier obtains solutions

(20)
$$u(x,t) = \frac{1}{x} \exp(-k\lambda^2 t)\sin \lambda x$$

provided the parameter λ is a solution of the transcendental equation

(21) $\dfrac{\lambda r}{tg\ \lambda r} = 1-hr.$

He easily proves that the equation has an infinity of real

roots λ_n tending to $+\infty$. To obtain a solution of (18) with

boundary condition (19) and such that $u(x,0)$ is a given

function $f(x)$, he proceeds as before, writing $xf(x)$ as a

series $\sum\limits_{n=1}^{\infty} c_n \sin \lambda_n x$; he shows that one has again the

"orthogonality" relations (of course he does not use that

word)

(22) $\displaystyle\int_0^r \sin \lambda_n x \sin \lambda_m x\ dx = 0$ for $m \neq n$

and from them deduces the relations

(23) $c_n = (\displaystyle\int_0^r xf(x)\sin \lambda_n x\ dx)/(\int_0^r \sin^2\lambda_n x\ dx)$

without of course any rigorous justification, nor any proof

of the fact that the series converges to $xf(x)$.

§3. The Sturm-Liouville theory.

The results of Fourier on the theory of heat were continued

and expanded by Poisson. Their work led Ch. Sturm in 1836

and J. Liouville one year later to build a general theory

which would include all cases considered by Fourier and

Poisson, without assuming the possibility of explicit inte-

gration. They consider a second order differential equation

(24) $y'' - q(x)y + \lambda y = 0$

where   q   is a real valued continuous function in a compact
interval [a,b] of $\mathbb{R}$, and λ a complex parameter. The
first problem is to consider boundary conditions of the form

$$(25)\quad y(a)\cos \alpha - y'(a)\sin \alpha = 0, \quad y(b)\cos \beta - y'(b)\sin \beta = 0$$

where α and β are two positive constants, and to determine
for what values of λ the problem has a non trivial solution
(an "eigenfunction" for the "eigenvalue" λ in our present
day language).

A first remark, which had already essentially been made by
Poisson, is that if λ, μ are two different eigenvalues, and
u, v two corresponding "eigenfunctions", then from the relations

$$u'' - qu + \lambda u = 0, \qquad v'' - qv + \mu v = 0,$$

one deduces

$$u''v - v''u + (\lambda-\mu)uv = 0$$

and as $\int_a^b (u''v-v''u)dx = (u'v-v'u)\Big|_a^b = 0$ because of (25),
one obtains

$$(26)\qquad\qquad (\lambda-\mu) \int_a^b u(x)v(x)dx = 0.$$

A first consequence of this relation is that eigenvalues are
necessarily <u>real</u> numbers. Indeed, if λ was not real, then
$\bar{\lambda}$ would also be an eigenvalue with eigenfunction $\bar{u}$, and
substituting $\bar{\lambda}$ and $\bar{u}$ for μ and v in (26), one obtains
$\int_a^b |u(x)|^2 dx = 0$, contrary to assumption.

The main contribution of Sturm was the proof that there are
infinitely many eigenvalues $\lambda_1 < \lambda_2 < ... < \lambda_n < ...$,   tending

to +∞. In his study of vibrating strings, d'Alembert had already considered an equation of the form $y'' - \lambda\varphi(x)y = 0$
where φ is not constant, and had tried to prove that there
is a single value of λ for which there is a solution in
$[a,b]$ vanishing at a and b and nowhere else; his idea
was to study the corresponding Riccati equation for y'/y
when λ varies $[65, (2), \text{vol. XI}_2, \text{p.311}]$. Sturm elects a
similar approach: he considers a solution $u(x,\lambda)$ of (24)
satisfying the <u>first</u> condition (25), and fixed for instance
by the condition $u'(a,\lambda) = 1$ (or $u'(a,\lambda) = 1$ if $\alpha = 0$),
and he studies the <u>variation</u> of $u(x,\lambda)$ as a function of λ;
the λ_n are therefore the solutions of the equation
$u(b,\lambda)\cos \beta - u'(b,\lambda)\sin \beta = 0$. He is thus led to <u>compare</u>
solutions of two equations

$$(27) \qquad y'' + q_1(x)y = 0, \qquad y'' + q_2(x)y = 0$$

when $q_1(x) \le q_2(x)$, and discovers many remarkable such
"comparison theorems", of which we will only quote the one
which leads to the existence of the eigenvalues. Sturm's
paper is rather long-winded and not very clear ($[209]$, $[S,$
p.259-268]) and there is a much simpler formulation of his
result: an equation $y'' + q(x)y = 0$ is written as a system
of two first order equations by the usual introduction of two
functions $y_1 = y$, $y_2 = y'$, which gives $y_1' = y_2$, $y_2' =$
$= -q(x)y_1$, and then one takes as new unknowns two functions
r,θ such that $y_1 = r \sin \theta$, $y_2 = r \cos \theta$, which leads to
the system

$$(28) \qquad r' = (1-q(x))r \sin \theta \cos \theta$$

$$(29) \qquad \theta' = \cos^2\theta + q(x)\sin^2\theta$$

where the second equation now is of the first order only[*].
The comparison theorem which is needed is then the following
one: consider solutions φ_1, φ_2 in $[a,b]$ of the two equations

$$(30) \quad \theta' = \cos^2\theta + q_1(x)\sin^2\theta, \qquad \theta' = \cos^2\theta + q_2(x)\sin^2\theta$$

and suppose that $q_1(x) < q_2(x)$ in $[a,b]$. Then, if for a
number $\alpha \in]a,b[$ one has $\varphi_1(\alpha) \le \varphi_2(\alpha)$, one also has
$\varphi_1(x) < \varphi_2(x)$ for $\alpha < x < b$. The proof is very simple and
consists in computing the derivative of the function $w(x) =$
$= \varphi_2(x) - \varphi_1(x)$ and showing that there is a continuous function f in $[a,b]$ such that $w'(x) - f(x)w(x) \ge 0$, which
implies that w cannot change sign.

If now we apply the preceding change of variable to (24), we
get the equation

$$(31) \qquad \theta' = \cos^2\theta + (\lambda - q(x))\sin^2\theta$$

and we consider the solution $w(x,\lambda)$ such that $w(a,\lambda) = \alpha$;
the eigenvalues λ are the solutions of the equations

$$(32) \qquad w(b,\lambda) = \beta + n\pi \qquad \text{for} \quad n \in \mathbb{Z}.$$

Sturm's comparison theorem then shows that for each $x \in]a,b]$
the function $\lambda \mapsto w(x,\lambda)$ is <u>strictly increasing</u>, and in addition, from (31) it follows that if $w(x,\lambda) = k\pi$ for an integer k, then $\frac{\partial w}{\partial x}(x,\lambda) = 1$. From these facts it is easy

---

[*]This device seems to have first been introduced by H. Prüfer
[180].

to show that each equation (32) has one and only one solution λ_n for each $n \geq 1$ and no solution for $n \leq 0$; in addition, the corresponding eigenfunction u_n may be shown to have exactly n zeroes in the interval $]a,b[$ [52, p.435-441].

Building on these results of Sturm, Liouville then proceeds to give a general formulation to the expansions of Fourier and Poisson. From relation (26) where λ and μ are replaced by λ_n and λ_m it follows that

$$(33) \qquad \int_a^b u_m(x)u_n(x)dx = 0 \qquad \text{for} \qquad m \neq n.$$

To each function f, defined and continuous in $[a,b]$, Liouville associates its "generalized Fourier coefficients"

$$(34) \qquad c_n = (\int_a^b f(x)u_n(x)dx)/(\int_a^b u_n^2(x)dx)$$

and considers the "generalized Fourier series" $\sum_{n=1}^{\infty} c_n u_n(x)$. In order to study its convergence, he needs more information on the behavior of λ_n and u_n when n tends to $+\infty$. He observes that, if $\lambda = \rho^2 > 0$, any solution of (24) satisfies a relation of the form

$$(35) \quad y(x) = A \cos \rho x + B \sin \rho x + \frac{1}{\rho} \int_a^x q(t)y(t)\sin \rho(x-t)dt$$

(which can be deduced from Lagrange's "variation of constants" method, by writing (24) as $y'' + \rho^2 y = q(x)y$, although this is not the way Liouville proves (35)). Applying this to $y = u_n$, so that ρ is replaced by $\lambda_n^{\frac{1}{2}}$, he gives a sketchy proof that $\rho_n = \frac{(n-1)\pi}{b-a} + O(1/n)$ and (if $\cos \alpha \neq 0$) $u_n(x) = \sqrt{\frac{2}{b-a}} \cos \rho_n x + O(1/n)$ (when u_n is normalized by

the condition $\int_a^b u_n^2(x)dx = 1$). This allows him to prove that
the series $\sum\limits_{n=1}^{\infty} c_n u_n(x)$ converges, provided the usual Fourier
series of f converges. He still has to show that, if f
is continuous, the function $F(x) = \sum\limits_{n=1}^{\infty} c_n u_n(x)$ is equal to
$f(x)$; he assumes (without proof) that F is continuous and
that $c_n = \int_a^b F(x)u_n(x)dx$, and is reduced to proving that
the relations $\int_a^b (F(x)-f(x))u_n(x)dx = 0$ for all n imply
$F = f$ (first appearance of the property of "completeness" of
an orthonormal system); but this he can only do under the additional assumption that $F-f$ has only a finite number of
zeroes in $[a,b]$. The complete proof of the relation $f(x) =$
$= \sum\limits_{n=1}^{\infty} c_n u_n(x)$ was only given (for f piecewise C^2) at the
end of the XIXth century, as well as the relation $\sum\limits_{n=1}^{\infty} c_n^2 =$
$= \int_a^b f^2(x)dx$; Liouville had only proved the corresponding
inequality $c_1^2 + \ldots + c_N^2 \leq \int_a^b f^2(x)dx$ for all N (named
after Bessel, who had proved it for the trigonometric system)
([151], [S, p.268-281]).

These remarkable results were to form the pattern of
spectral theory, the main efforts of analysts in that direction being directed to a generalization of the Sturm-Liouville
theory to some types of partial differential equations; but
in the first half of the XIXth century, the theory of these
equations was far less advanced than the theory of ordinary
differential equations, and it is only after 1880 that progress
became possible (see Chapter III).

CHAPTER II

THE "CRYPTO-INTEGRAL" EQUATIONS

§1. The method of successive approximations.

The study of celestial mechanics during the XVIIIth century
by the method of perturbations consisted, for the theory of
the movements of planets, to first neglect their mutual attraction, which gave for each planet a Keplerian orbit around
the sun, and then to find the deviations of the actual orbits
from the Keplerian ones by taking into account the attraction
of other planets; due to the fact that the masses of the planets are much smaller than the mass of the sun, these deviations
were expected to be small. Translated into mathematical terms,
this amounted, in the simplest cases, to find good approximations
for the solutions of a system of differential equations

$$(1) \quad y_i' = \varepsilon f_{1i}(x, y_1, \ldots, y_n) + \varepsilon^2 f_{2i}(x, y_1, \ldots, y_n) + \ldots \quad (1 \le i \le n)$$

where the parameter ε on the right-hand sides is "small".
The general conception of function in XVIIIth century mathematics naturally led to try to express the y_i as a power
series in ε

$$(2) \quad y_i = a_i + \varepsilon y_{1i} + \varepsilon^2 y_{2i} + \ldots \quad (1 \le i \le n),$$

22

to substitute these expressions in (1) and identify the coefficients of the successive powers of ε on both sides. This
led to a succession of equations

$$y'_{1i} = f_{1i}(x, a_1, \ldots, a_n)$$

$$y'_{2i} = F_{2i}(x, y_{11}, \ldots, y_{1n})$$

$$y'_{3i} = F_{3i}(x, y_{11}, \ldots, y_{1n}, y_{21}, \ldots, y_{2n})$$

. . . .

all of which had right-hand sides which were known functions,
hence were reduced to mere "quadratures". No attempt was made
to justify mathematically those procedures; the goal of these
computations was merely to obtain a satisfactory agreement
with observations.

It is well-known that Cauchy was the first mathematician who
proved existence theorems for general types of differential
equations, for which no explicit solution is available. His
strategy was to consider the various methods introduced earlier
for the purpose of numerical computations, and to show that,
under certain conditions, these methods actually gave convergent approximation processes having a solution as limit.
In particular, in a paper published in 1835 in Prag ([40],(2),
vol. XI, p.399-465), he takes up the method outlined above,
not for an ordinary differential equation, but for a linear
partial differential equation of first order (which was known
to be equivalent to a system of ordinary differential equations)

(3) $$\frac{\partial U}{\partial t} = \sum_{i=1}^{p} A_i(t, x_1, \ldots, x_p) \frac{\partial U}{\partial x_i};$$

the problem is to find a solution which for t = 0 reduces
to a given function $u(x_1, \ldots, x_n)$, and Cauchy transforms (3)
into the equivalent "integro-differential" equation by considering $x_1, \ldots, x_p$ as parameters:

$$(4) \quad U(t, x_1, \ldots, x_p) = u(x_1, \ldots, x_p) + \int_0^t \left(\sum_{i=1}^p A_i(s, x_1, \ldots, x_p) \frac{\partial U}{\partial x_i} \right) ds$$

which he solves by successive approximations, starting with
$U_o = u$, and defining

$$U_n(t, x_1, \ldots, x_p) = u(x_1, \ldots, x_p) + \int_0^t \left(\sum_{i=1}^p A_i(s, x_1, \ldots, x_p) \frac{\partial U_{n-1}}{\partial x_i} \right) ds$$

by induction; but he is only able to prove convergence towards
a solution when the A_i are analytic functions.

  In his 1837 papers on the Sturm-Liouville problem Liouville
independently applied a similar method to the linear differential equation $y'' = f(x)y$, for which he wants to find a
solution in $[a,b]$ satisfying the boundary condition $y'(a)$ -
- $hy(a) = 0$. He starts from the function $y_o(x) = 1+h(x-a)$
satisfying that condition, and considers the series

$$(5) \qquad\qquad y = y_o + y_1 + \ldots + y_n + \ldots$$

where the y_n are determined for n > 0 by the recursive
equations

$$y_{n+1}(x) = \int_a^x dt \int_a^t f(s) y_n(s) ds.$$

It must be remembered that at that time the concept of uniform convergence had not yet been formulated, and no justification had been given for asserting the continuity of a convergent series of continuous functions, or differentiating or

integrating such a series termwise. Liouville proves very

easily that there is a constant C such that

$$|y_n(x)| \le c^n(x-a)^{2n}/(2n)!$$

from which he concludes that the series (5) giving $y(x)$ con-

verges for every x; but he tacitly takes for granted that y

is a c^2 function and a solution of his problem.

 In addition, Liouville makes the interesting remark that the

function y can also be defined by the relation

(6) $$y = y_o + \int_a^x dt \int_a^t f(s)y(s)ds$$

(which he could also have written $y = y_o + \int_a^x (x-t)f(t)y(t)dt)$,

thus giving what is probably the first example of what will

be called later a "Volterra integral equation of the second

kind" (see chap. IV); if one writes $z_n = y_o + y_1 + \ldots + y_n$,

Liouville observes that the z_n are given by $z_o = y_o$ and

the recursive equations

(7) $$z_{n+1}(x) = y_o + \int_a^x dt \int_a^t f(s)z_n(s)ds$$

which is the standard process of "successive approximations"

for these equations ([151], [S, p.268-281]).

 We have already seen that a little later in his papers of

1837, Liouville gives another "integral equation" equivalent

to an equation $y'' = f(x)y$ (chap. I, §3, equation (35)).

This exemplifies a general idea: if a linear differential

operator P is such that the equation $P \cdot u = f$ can be sol-

ved by a formula $u = y_o + G \cdot f$, where G is a linear operator,

then the equation $P \cdot u + Q \cdot u = 0$, where Q is an operator,
is equivalent to $u - G \cdot (Q \cdot u) = y_0$; in the case of Liouville,
$P \cdot u = u'' + \rho^2 u$ and $Q \cdot u = -qu$, and G is an integral operator (cf. chap. IX, §5).

The simplest application of this idea is to the proof of
Cauchy's existence and uniqueness theorem for an ordinary differential equation $y' = f(x,y)$, which, with the initial condition $y(x_0) = y_0$, is equivalent to $y = y_0 + \int_{x_0}^{x} f(t,y)dt$.
In this general form it is given by E. Picard in his 1890
paper on successive approximations [172, vol. II, p.197-200],
where it comes as an afterthought, the bulk of the paper being
concerned with applications of the method to partial differential equations. However, in these applications, Picard is
directly influenced by the fundamental earlier works of
C. Neumann on the Laplace equation and of H.A. Schwarz on the
equation of vibrating membranes, which are the direct forerun-
ners of the theory of integral equations; we will describe in
detail C. Neumann's results in §5 of this chapter, and H.A.
Schwarz's paper in chap. III, §1.

§2. Partial differential equations in the XIXth century.

During the whole XIXth century, the theory of partial differential equations (in contrast with the theory of ordinary
differential equations) has remained in an embryonic stage.
The only general theorem, patterned after the Cauchy theorem
on local existence and uniqueness of solutions of ordinary
differential equations, is the Cauchy-Kowalewska theorem:
suppose we have a system of r equations in r unknown real

functions $v_1, \ldots, v_r$ of $p+1$ real variables $x_1, \ldots, x_{p+1}$, of type

$$\frac{\partial v_j}{\partial x_{p+1}} =$$

(8)

$$= H_j(x_1, \ldots, x_{p+1}, v_1, \ldots, v_r, \frac{\partial v_1}{\partial x_1}, \frac{\partial v_1}{\partial x_2}, \ldots, \frac{\partial v_r}{\partial x_{p-1}}, \frac{\partial v_r}{\partial x_p}) \quad (1 \le j \le r)$$

where the right hand sides do not contain any derivative with respect to x_{p+1}, and are supposed to be real and <u>analytic</u> with respect to their $p+1+r+rp$ variables, in a neighborhood V_0 of 0 in $\mathbb{R}^{p+1+r+rp}$; then there is a small neighborhood V of 0 in $\mathbb{R}^{p+1}$ such that (8) has in V a unique solution $(v_1, \ldots, v_r)$ consisting of <u>analytic</u> functions in V, such that $v_j(x_1, \ldots, x_p, 0) = 0$ in $V \cap \mathbb{R}^p$ for $1 \le j \le r$.

The tendency (inherited from the XVIIIth century) to consider that the most interesting functions were analytic was still very strong during the whole XIXth century, and therefore at first the analyticity restrictions of the Cauchy-Kowalewska theorem did not worry mathematicians very much. However, as it was known that some special types of partial differential equations, such as the scalar equation of first order and some types of second order equations, had solutions under much less stringent restrictions, people began to wonder if some other method than Cauchy's "method of majorants" (which could only be applied to analytic functions) would not yield a generalization of the Cauchy-Kowalewska theorem, at least for C^∞ functions. The question remained unanswered until 1956, when H. Lewy gave the surprising example of a system of two linear equations in 3 variables, with C^∞ coefficients

$$\begin{cases} \dfrac{\partial v_1}{\partial x_1} = \dfrac{\partial v_2}{\partial x_2} - 2x_2 \dfrac{\partial v_1}{\partial x_3} - 2x_1 \dfrac{\partial v_2}{\partial x_3} - f(x_3) \\[3mm] \dfrac{\partial v_2}{\partial x_1} = - \dfrac{\partial v_1}{\partial x_2} + 2x_1 \dfrac{\partial v_1}{\partial x_3} - 2x_2 \dfrac{\partial v_2}{\partial x_3} \end{cases}$$

which, for a suitable choice of the real C^∞ function f,
has no solution whatsoever around any point (even if one allows solutions which are distributions).

We shall not discuss the numerous local studies of analytic
systems of partial differential equations (not necessarily
reducible to the form (8)) which followed the Cauchy-Kowalewska
theorem, since they had no influence on the development of
Functional Analysis as we understand it.

The remainder of the theory of partial differential equations until 1890 was limited to very special scalar equations
(mostly linear equations or order 2) generally derived from
physical problems[*], such as the equation of vibrating strings
and its generalizations to 3 and 4 variables (the "wave equations"), the Laplace equation $\Delta u = 0$ in 2 and 3 variables,
the heat equation in 2, 3 and 4 variables. For these equations, the techniques of "separation of variables" or of
Fourier transforms (see chapter VII, §6) gave special solutions or solutions depending on "arbitrary" functions. But
until 1825 the determination of solutions by boundary conditions (of which we have seen a few examples in Chapter I) was
always restricted to explicitly described and particular such

---

[*] See the interesting description of these problems given by
Poincaré in the Introduction of his 1890 paper on the
equations of mathematical physics ([177], vol.IX, p.28-32)

conditions.

A first attempt of classification of second order equations
in 2 variables had been made by Laplace [137, vol. IX, p.21-28].
He considered "quasi-linear" equations, i.e. those of the form

$$A(x,y) \frac{\partial^2 z}{\partial x^2} + B(x,y) \frac{\partial^2 z}{\partial x \partial y} +$$

(9)

$$+ C(x,y) \frac{\partial^2 z}{\partial y^2} + F(x,y,z, \frac{\partial z}{\partial x}, \frac{\partial z}{\partial y}) = 0$$

linear in the second order derivatives. As he did not have a
clear idea of the distinction between real and complex variables, and therefore did not hesitate to give complex values
to x and y, he asserted that a suitable change of variables could reduce the terms of (9) containing second order
derivatives either to $\frac{\partial^2 z}{\partial x \partial y}$ or to $\frac{\partial^2 z}{\partial x^2}$ when A, B, C are
not all identically zero! With the development of the theory
of functions of one complex variable, it was soon realized
that, for real variables x, y, equations (9) where the second
order derivatives enter by $\frac{\partial^2 z}{\partial x^2} + \frac{\partial^2 z}{\partial y^2}$ (called _elliptic_ equations) had to be sharply distinguished from those (called
hyperbolic equations.) where the second order derivatives enter
by $\frac{\partial^2 z}{\partial x \partial y}$ or $\frac{\partial^2 z}{\partial x^2} - \frac{\partial^2 z}{\partial y^2}$. The study of general boundary conditions for hyperbolic equations only begins around 1860 and
will have little contact with Functional Analysis until around
1925 (see chapter IX, §5). On the contrary, the various problems connected with the Laplace equation in 2 or 3 variables
will be one of the main concerns of analysts from 1828 onwards,
and will become the impetus leading to the theory of integral
equations, and thence to our modern Functional Analysis.

§3. The beginnings of potential theory.

In 1748, D. Bernoulli had introduced in the theory of newtonian attraction the function $\Omega(M) = \sum_i (m_i\mu/r_i)$ for a point
M of mass μ attracted by a finite number of punctual masses m_i, where r_i is the distance of M to the mass m_i;
and in 1773 Lagrange observed that the knowledge of that function immediately gave the components of the attraction exerted
on M, by taking the derivatives of Ω with respect to the
coordinates x, y, z of M. When the finite number of masses
is replaced by a solid V of density ρ and the point M is
outside V, the function Ω becomes

(10) $$\Omega(x,y,z) = \mu \iiint_V \frac{\rho(\xi,\eta,\zeta)d\xi d\eta d\zeta}{r(x,y,z,\xi,\eta,\zeta)}$$

with $r(x,y,z,\xi,\eta,\zeta) = ((x-\xi)^2 + (y-\eta)^2 + (z-\zeta)^2)^{\frac{1}{2}}$ [135, vol.
VI, p.349].

In 1782 and 1785, Laplace showed that outside of V the function Ω satisfied the equation

(11) $$\Delta\Omega \equiv \frac{\partial^2\Omega}{\partial x^2} + \frac{\partial^2\Omega}{\partial y^2} + \frac{\partial^2\Omega}{\partial z^2} = 0$$

[137, vol. X, p.361-363 and vol. XI, p.276-280], and in 1813
Poisson completed that result by showing that if ρ is con-
tinuous in V, the integral (10) is still meaningful inside
V, and Ω satisfies the "Poisson equation"

(12) $$\Delta\Omega + 4\pi\rho = 0$$

([178], [S, p.342-346]). His idea is to consider the value
of Ω at a point M in V as the sum of the corresponding

functions Ω_1, Ω_2 relative to a small ball V_1 of center M
and to the complement V_2 of V_1 in V; one has then
$\Delta\Omega_2 = 0$, and when the radius of V_1 tends to 0 Poisson
shows that $\Delta\Omega_1$ tends to $-4\pi\rho(M)$. (In fact his argument is
not rigorous when one only assumes the continuity of ρ, and
the existence of $\Delta\Omega$ is only guaranteed when ρ satisfies a
Hölder condition; when ρ is merely continuous, equation (12)
is valid only if the second order derivatives are taken in
the sense of the theory of distributions (chap. VIII, §3)).

After the discovery of Coulomb's laws (1785) the Laplace
equation became of central importance in electrostatics; it
also was found to govern "stationary" phenomena in hydro-
dynamics and the theory of heat. Finally the so-called
"Cauchy-Riemann" equations for real functions P, Q of x, y
such that P + iQ is an analytic function of x + iy, were
known since the middle of the XVIIIth century, and they
implied that P and Q were solutions of the Laplace equation in 2 variables. Very early in the XIXth century, Gauss
was well aware of this connection and of the fact that one
obtained solutions of the Laplace equation in 2 variables by
replacing the function (10) by

$$(13) \qquad \Omega(x,y) = \iint_D \rho(\xi,\eta) \log \frac{1}{r(x,y,\xi,\eta)} \, d\xi d\eta$$

for a bounded domain D in the plane. The development by
Cauchy of the theory of holomorphic functions of a complex
variable could thus be used to yield properties of harmonic
functions of 2 variables, such as for instance the non existence of relative extrema for such a function in its domain

of definition; it was then natural to conjecture that similar
properties were also valid for harmonic functions of 3 (and
later for $n \geq 4$) variables, although they had to be proved
by other means.

The first paper dealing with <u>general</u> boundary conditions for
a partial differential equation was written in 1828 by George
Green, a self-taught English mathematician (1793-1841); it is
concerned with electrostatics and the general study in that
theory of what Green for the first time calls <u>potential func-</u>
<u>tions</u>. By that he not only means the functions of the form
(10), but also what will later be called <u>simple layer poten-</u>
<u>tials</u>, namely functions of the type

$$(14) \qquad \Omega(M) = \iint_\Sigma \frac{\rho(P)}{MP} \, d\sigma(P)$$

where Σ is a smooth surface, ρ (the "density") a continuous function on Σ and $d\sigma$ the element of area on Σ; he
was naturally led to such functions by the known experimental
fact that on conductors the electric charges are concentrated
on their surface.

Green was interested in the relations between the surface
density ρ and the potential it defines. He first estab-
lishes the famous theorem which, for the operator Δ, generalizes to 3 dimensions the relation between a differential
operator and its adjoint (Chapter I, formula (5)):

$$(15) \qquad \iiint_V (u\Delta v - v\Delta u)d\omega = \iint_\Sigma (v\frac{\partial u}{\partial n} - u\frac{\partial v}{\partial n})d\sigma$$

where Σ is a smooth surface limiting a bounded volume V,
u and v are C^2 in a neighborhood of $\bar{V}$, $\frac{\partial u}{\partial n}$ is the de-

rivative of u along the exterior normal of Σ [*]. He then
has the original idea [**] of considering a function u which,
still C^2 for all points different from a point M in V,
becomes infinite at M in such a way that the difference
$u(P) - (1/MP)$ is bounded when P tends to M; he applies
(15) to the volume V from which a small ball of center M
has been excised, and by letting the radius of the ball tend
to 0, he obtains the formula

$$(16) \quad 4\pi v(M) + \iiint_V (u\Delta v - v\Delta u)d\omega = \iint_\Sigma (v\frac{\partial u}{\partial n} - u\frac{\partial v}{\partial n})d\sigma$$

provided of course the triple integral exists. Taking in particular $u(P) = 1/MP$ would give for a solution v of $\Delta v = 0$

$$(17) \quad 4\pi v(M) = \iint_\Sigma (v\frac{\partial(\frac{1}{r})}{\partial n} - \frac{1}{r}\frac{\partial v}{\partial n})d\sigma \qquad (\text{with} \quad r(P) = MP)$$

in other words, an integral formula which would solve the
Laplace equation when v and $\frac{\partial v}{\partial n}$ were known on Σ. This
was in agreement with what was known at the time for partial
differential equations of the second order, such as the equation of vibrating strings (Chapter I, §2). However, experi-
ments showed that v was entirely determined by its values
on Σ, and therefore it was not possible to take for both v

---

[*]Lagrange [135, vol.I, p.263] and Gauss [82,vol.V,p.22] had
    already obtained more particular relations of that kind
between volume and surface integrals.

[**]It is of course the same idea which leads to the Cauchy
    formula giving the value of a holomorphic function inside
a domain D when it is known on the boundary of D. However,
it is unlikely that Green knew Cauchy's papers

and $\dfrac{\partial v}{\partial n}$ on Σ <u>arbitrary</u> continuous functions, so that the situation appeared quite different from the boundary conditions for hyperbolic equations. Furthermore, there was at least one case when an explicit formula gave v inside V by an integral extended to Σ, namely the <u>Poisson formula</u> for a ball V of center O and radius a, published in 1820:

$$(18)\qquad v(M) = \frac{1}{4\pi} \iint_{\Sigma} \frac{a^2 - \rho^2}{ar^3} v(P) d\sigma$$

with $\rho = OM$. Green observed that one would have a similar formula for general domains V:

$$(19)\qquad v(M) = \frac{1}{4\pi} \iint_{\Sigma} v(P) \frac{\partial G}{\partial n}(M,P) d\sigma$$

by substituting in his formula (16) for u a function $G(M,P)$ such that: $1^{\underline{o}}$ in $V \times V$, G is C^2 provided $M \neq P$ and $\dfrac{\partial G}{\partial n}$ exists on Σ; $2^{\underline{o}}$ $G(M,P) - 1/MP$ remains bounded when P tends to M; $3^{\underline{o}}$ $G(M,P) = 0$ when M is in V and P on Σ; $4^{\underline{o}}$ when M is fixed in V, G, as a function of P, satisfies the Laplace equation in V. He could not prove the existence of such a "Green function", but made it plausible by an appeal to experimental facts: when the surface Σ is connected to the ground, and an electric charge $+1$ is put at the point M, it "induces" an electric charge on Σ such that the total potential of that charge and the punctual charge at M is 0 on Σ; that potential should be the function $G(M,P)$ ([90], [S, p.347-358]).

Finally, by an ingenious use of his formula (15), Green could prove that in $V \times V$, one had $G(P,M) = G(M,P)$ for $M \neq P$.

§4. The Dirichlet principle.

Gauss had very early been interested in the Laplace equation, both in 2 variables in connection with his work on complex numbers, and in 3 variables in relation with his astronomical studies, and we have seen that in his 1813 paper on the attraction of spheroids, he had proved particular cases of the Green formula (15). After 1830, he devoted much of his time to the study of magnetism, both experimentally and theoretically, and thus was led to new research on potential theory, which he published in 1840 [82, vol. V, p.197-242]. In that paper, he quotes no other work on the subject, and it is very unlikely that he ever heard of Green (whose work was not widely known, even in England)[*]; he expands his 1813 formulas and obtains in this way some new particular cases of Green's formula (15), although he does not seem to have thought of formula (16). The closest approach to the latter is his famous "mean value formula"

$$(20) \qquad v(0) = \frac{1}{4\pi} \iint_{\Sigma} v(P) d\sigma$$

for a harmonic function v in a sphere Σ of center 0, for which it is quite surprising that he should not have observed that it was a special case of Poisson's formula (18) which he cannot have failed to know.

---

[*] The fact that Gauss also uses the word "potential" with the same meaning may be attributed to the fact that the word (in its Latin form) was commonly used in the XVIIIth century by "natural philosophers".

As Green had done, Gauss was particularly interested in the behavior of simple layer potentials (14) when M tends to a point on the surface Σ; by a careful study, he shows that the potential Ω is continuous everywhere, and that the normal derivatives at a point M_o of Σ exist on both sides of the surface, but have <u>different values</u>, their difference being $4\pi\rho(M_o)$; all this had been taken for granted without proof by Green.

Gauss attacked several problems related to potential theory, some of which were to become the focus of active research after 1930. One was the <u>equilibrium problem</u>: find a distribution of electric charges on a closed surface Σ giving a potential which is <u>constant</u> on Σ; another consisted in replacing charges inside Σ by charges on Σ in such a way that the potential outside Σ remains the same (what would later be called a "sweeping-out" process), and Gauss showed that it could be solved if the equilibrium problem had a solution.

Regarding the latter, Gauss introduced a new idea which was to become quite central in potential theory: he observed that if the potential Ω is given by (14) with $\rho \geq 0$, and U is any continuous function on Σ, then if ρ is chosen such that the integral $\iint_{\Sigma} (\Omega - 2U)\rho\,d\sigma$ takes the <u>smallest</u> possible value among all possible choices of ρ, then $\Omega - U$ is constant on Σ, and he added that the existence of such a density ρ was obvious.

By adding to Ω a suitable constant, this method of Gauss solved the problem of finding a harmonic function u in the volume V, continuous in $\bar{V} = V \cup \Sigma$, and equal on Σ to a given function U $^{(*)}$. The same problem was considered a little later by W. Thompson (the future Lord Kelvin) in 1847 and by Dirichlet around the same time in his lectures (published long afterwards) [S, p.380-387]; it became known as the Dirichlet problem. Their idea is similar to Gauss's: they consider the volume integral

$$(21) \qquad \iiint_V \left(\left(\frac{\partial v}{\partial x}\right)^2 + \left(\frac{\partial v}{\partial y}\right)^2 + \left(\frac{\partial v}{\partial z}\right)^2 \right) d\omega$$

and the function v continuous in $\bar{V}$ with continuous and bounded first derivatives in V (v taking the given values on Σ), for which the integral (21) takes its smallest value; applying the standard techniques of the Calculus of variations, they easily show that such a function is indeed harmonic in V. The great success of this idea is probably due to the imaginative use Riemann almost immediately made of it, in his epoch-making papers on holomorphic functions, Riemann surfaces and abelian integrals. By considering the real and imaginary parts of such functions, he was the first to realize that the existence theorems he needed could be derived from similar existence theorems for these harmonic functions, which he thought he could prove by adapting Dirichlet's argument to similar integrals in 2 variables, called by him "Dirichlet principle" [182, p.97].

$^{(*)}$If such a problem is solved, it implies the existence of the Green function: one considers the function $u(M,P)$ harmonic in V (as a function of P) which takes the values $-1/MP$ on Σ; the Green function is then $G(M,P) = u(M,P)+(1/MP)$, provided one shows that $\frac{\partial G}{\partial n}$ exists and is continuous on Σ.

His magnificent results attracted considerable attention,
but soon mathematicians realized that they rested on three
properties for which W. Thompson, Dirichlet and Riemann did
not give any proof at all:

1) For a given continuous function g on Σ,   there exist
continuous functions v in $\bar{V}$ whose restriction to Σ is
g and for which the integral (21) is meaningful.

2) If such functions exist, there is one for which the
smallest value of (21) is attained.

3) For that function v,   the second order derivatives
$\frac{\partial^2 v}{\partial x^2}$, $\frac{\partial^2 v}{\partial y^2}$, $\frac{\partial^2 v}{\partial z^2}$ exist.

However, in 1871, F. Prym presented an example (for two variables and V the disk $x^2 + y^2 < 1$) where no function v
satisfying 1) existed [181][*]. On the other hand, in 1870,
Weierstrass observed that in all problems of the Calculus of
variations which had been studied since the beginning of the
XVIIIth century, properties 2) and 3) had been taken for
granted without any proof, and he gave a very simple example
in which property 2) does not hold: the problem of minimizing
the integral $\int_{-1}^{1} xy'^2 dx$ among all C^1 functions y defined
in the interval [-1,1] and satisfying the boundary condi-
tions $y(-1) = a$, $y(1) = b$, with $a \neq b$ [S, p.390-391].
Spurred by these difficulties, Weierstrass and his pupils
(P. Du Bois-Reymond, A. Kneser, S. Zaremba) undertook to put

(*)
  The discovery of that fact is usually attributed to Hadamard, who published a similar example in 1906 [94,vol.III,
  p. 1245-1248].

the Calculus of variations on sounder foundations and were
able to rescue many classical results from the suspicion
raised by such counterexamples. But the "Dirichlet principle"
eluded their efforts, and it was only in 1899 that Hilbert,
using new ideas in what was called his "direct method", was
able to give a complete justification of the use Riemann had
made of that "principle" [111, vol. III, p.10-37].

§5. The Beer-Neumann method.

We shall see in later chapters how the concepts and tools
used by Hilbert and the Weierstrass school contributed to the
birth of General Topology and later to the introduction of
such notions as "weak" solutions of partial differential
equations. Meanwhile, the challenge remained to prove the
existence of a solution to the Dirichlet problem and similar
boundary values problems for the Laplace equation, at least
under conditions such as were used in Riemann's work. Between 1870 and 1890, that challenge was successfully taken up
by three mathematicians: H.A. Schwarz around 1870, C. Neumann
in 1877 and H. Poincaré in 1887.

We shall not discuss in detail the contributions of Schwarz
and Poincaré, which did not influence directly the development of Functional Analysis. Both are based on the idea of
approximation: starting from known solutions of the Dirichlet
problem for special kinds of domains, an approximation process
enables one to get solutions for much more general domains.
Schwarz limits himself to 2 variables; he first considers
domains limited by a convex polygon, for which it is possible

to prove directly (by explicit construction) the existence of
a conformal mapping on the unit disk, hence the existence of
a solution of the Dirichlet problem (by transferring the
Poisson formula from the circle to the polygon). Using the
maximum principle, it is then possible to prove the existence
of the solution for a <u>convex</u> domain by approximating it by a
sequence of inscribed convex polygons. A little later, he
invented an ingenious "alternating process" which enabled him
to show that when one can solve the Dirichlet problem for two
domains in the plane, it is also possible to solve it for their
union, and from that result he finally showed that the
Dirichlet problem in the plane is solvable for any domain
limited by piecewise analytic curves [196, vol.II, p.133-210].

Poincaré's famous "sweeping-out method" applies to any number
of dimensions. To solve the Dirichlet problem for a bounded
domain V limited by a surface Σ, he shows (using the maximum principle) that it is enough to consider the case in which
the function given on Σ is the restriction to Σ of a function Φ defined in a neighborhood W of $\bar{V}$, of class C^2
and such that $\Delta\Phi \geq 0$. By Poisson's equation (12), Φ is
the sum of a harmonic function and a potential Φ_o of masses
≥ 0. The fundamental idea is that if B is a ball contained
in W, it is possible to use the Poisson integral (18) extended to the surface of B in order to replace Φ_o by another
potential which coincides with Φ_o outside B and is smaller
than Φ_o inside B; the masses inside B have been "swept
out" on the surface of B. One then takes an infinite sequence of balls B_n, whose union is V, and one applies the

"sweeping-out" process repeatedly to the B_n, in the order $B_1, B_2, B_1, B_2, B_3, B_1, B_2, B_3, B_4, \ldots$ (each B_n is "swept-out" in-finitely many times). The corresponding sequence of potent-ials is decreasing, hence has a limit in V; using Harnack's inequalities (consequences of the Poisson formula (18)) and the maximum principle, Poincaré is able to show that this lim-it is a solution of the Dirichlet problem, provided the boundary Σ satisfies a "regularity" condition, namely, for any point $M \in \Sigma$, there must be a small ball whose intersec-tion with $\bar{V}$ is reduced to M [177, vol. IX, p.33-54]; later, Zaremba could replace the small ball by a small cone of vertex M in that condition.

In contrast with Schwarz's and Poincaré's papers, the Beer-Neumann method was a landmark in Functional Analysis by in-troducing the first example of what was later to be called a "Fredholm integral equation of the second kind". Green's formula (17) naturally introduced still another type of po-tential:

$$(22) \qquad u(M) = \iint_{\Sigma} \rho(P) \frac{\partial}{\partial n} \left(\frac{1}{MP}\right) d\sigma$$

which was harmonic outside the surface Σ. It also occurred in the theory of magnetism, from which it got its name of double layer potential: it was there conceived as the limit of a difference of two simple layer potentials, one with den-sity μ on Σ, the other with density μ on a surface Σ' parallel to Σ and at an "infinitely small" distance ε; when ε tends to 0, μ was supposed to increase to $+\infty$ in such a way that the product $\mu\varepsilon$ tended to ρ.

Such a potential had been shown to have near Σ a behavior quite similar to the normal derivative of a simple layer potential, studied by Gauss: when M tends to a point M_o of Σ along the normal to Σ at M_o, $u(M)$ tends to a limit on each side of Σ, but these limits are <u>different</u> in general; however, $\frac{\partial u}{\partial n}$ is the <u>same</u> on both sides.

Formula (22) also had a nice geometric interpretation; one has $\frac{\partial}{\partial n} \left(\frac{1}{MP}\right) = \frac{\cos \varphi}{MP^2}$ where φ is the angle between MP with the normal to Σ at P, and $\frac{\cos \varphi}{MP^2} d\sigma$ is the infinitesimal "solid angle" from which $d\sigma$ is "seen" from the point M.

Around 1860, A. Beer proposed to obtain a solution to the Dirichlet problem by formula (22) for a suitable density ρ on Σ. From the continuity properties of double layer potentials, it follows that if Σ is a smooth surface, and $g(M)$ is the function on Σ to which the solution $u(M)$ must be equal, the unknown density must satisfy the equation

$$(23) \qquad 2\pi\rho(M) + \iint_{\Sigma} \rho(P) \frac{\partial}{\partial n} \left(\frac{1}{MP}\right) d\sigma = g(M) \qquad \text{for} \quad M \in \Sigma.$$

He then concluded that one could compute ρ by the usual device of "successive approximations" (§1) starting with $\rho_o(M) = \frac{1}{2\pi} g(M)$ and defining recursively $\rho_n(M)$ by

$$2\pi\rho_n(M) + \iint_{\Sigma} \rho_{n-1}(P) \frac{\partial}{\partial n} \left(\frac{1}{MP}\right) d\sigma = 0 \qquad \text{for} \quad n \geq 1$$

so that the series $\rho(M) = \rho_o(M) + \rho_1(M) + \ldots + \rho_n(M) + \ldots$ would give the solution to (23); but he made no attempt to prove that the series converged.

In 1877, Carl Neumann attempted to give such a proof [165].
He restricted himself to the case in which the domain   V   is
bounded and convex, but he allowed a non smooth boundary Σ;
equation (23) must then be modified to

$$(24) \qquad 4\pi\rho(M) = \iint_{\Sigma} (\rho(M)-\rho(P)) \frac{\cos\varphi}{MP^2} d\sigma + f(M)$$

with   f   continuous on Σ,   and the successive approximations
are given by $4\pi\rho_0(M) = f(M)$ and, for $n \geq 1$,

$$(25) \qquad 4\pi\rho_n(M) = \iint_{\Sigma} (\rho_{n-1}(M) - \rho_{n-1}(P)) \frac{\cos\varphi}{MP^2} d\sigma.$$

Neumann's idea is to consider the maximum value L_n and minimum value ℓ_n of ρ_n,   and to show that there is a number
q   such that $0 < q < 1$ and

$$(26) \qquad L_n - \ell_n \leq (L_0 - \ell_0)q^{n-1}$$

from which he majorizes $|\rho_n(M)|$ by a multiple of q^n using
(25), and he can conclude that the series $\sum_{n=0}^{\infty} \rho_n(M)$ converges to a continuous function.

To prove (26), Neumann divides Σ into two parts A_n, B_n
respectively defined by the conditions

$$\frac{1}{2}(L_{n-1} + \ell_{n-1}) \leq \rho_{n-1}(P) \leq L_{n-1} \qquad \text{for} \quad A_n$$

$$\ell_{n-1} \leq \rho_{n-1}(P) < \frac{1}{2}(L_{n-1} + \ell_{n-1}) \qquad \text{for} \quad B_n$$

and he deduces from (25) that for all points   M   of Σ

$$(L_{n-1}-\ell_{n-1})(A_n(M)+ \frac{1}{2}B_n(M)) \leq 4\pi\rho_n(M) \leq (L_{n-1}-\ell_{n-1})(\frac{1}{2}A_n(M)+B_n(M))$$

where $A_n(M)$ and $B_n(M)$ are the solid angles from which A_n and B_n are "seen" from M. This implies

$$L_n - \ell_n \le (L_{n-1} - \ell_{n-1})q$$

where q is the <u>least upper bound</u> of the quantity

$$(27) \quad \Lambda(M, M', A, B) = \frac{1}{4\pi} \left(\frac{1}{2} A(M) + B(M) + A(M') + \frac{1}{2} B(M') \right)$$

when M and M' vary arbitrarily in Σ, A is an arbitrary closed part of Σ and B its complement. One is thus faced with the <u>purely geometric</u> problem of showing that $q < 1$. The expression (27) can be written

$$\frac{1}{4\pi} \left(A(M) + B(M) + A(M') + B(M') - \frac{1}{2} (A(M) + B(M')) \right)$$

and also

$$\frac{1}{4\pi} \left(\frac{1}{2} (A(M) + B(M)) + \frac{1}{2} (A(M') + B(M')) + \frac{1}{2} (A(M') + B(M)) \right)$$

and as one always has $A(M) + B(M) \le 2\pi$ (maximum value of the solid angle from which the whole of Σ is "seen" from one of its points), the problem can also be formulated in two equivalent ways:

$$(28) \qquad A(M) + B(M') \ge 4\pi r \quad \text{for an} \quad r > 0,$$

$$(29) \qquad A(M) + B(M') \le 4\pi s \quad \text{for an} \quad s < 1,$$

for <u>all</u> points M, M' in Σ and a division in two parts A, B. The form (28) of that condition immediately shows that there is an <u>exceptional</u> type of convex set for which it cannot be satisfied, namely the case in which V is the <u>intersection of two</u>

convex cones ("double cone"): indeed we then have $A(M) =$
$= B(M') = 0$ if A is the surface of one of the cones, B
the surface of the other, M the vertex of A and M' the
vertex of B. Furthermore, this particular choice of A, B, M
and M' is the only one for which $A(M) + B(M')$ may be 0.
However, when the exceptional case is excluded, Neumann concludes, from the fact that $A(M) + B(M') > 0$ for all choices
of A, B, M and M', that there is an $r > 0$ for which (28)
is satisfied for all these choices, and does not give a proof
of that assertion valid for all convex sets other than double
cones. This gap in Neumann's proof seems to have remained
undetected until Lebesgue drew attention to it in 1937 ([138],
vol. IV, p.151-166). He shows in addition how one can fill
in that gap by a compactness argument: there are two points
M_o, M_o' in Σ, limits of sequences (M_k), (M_k') such that for
each k there is a splitting of Σ in two parts A_k, B_k,
such that $A_k(M_k) + B_k(M_k')$ tends to the l.u.b. $4\pi s$ of
$A(M) + B(M')$ for all choices of A, B, M, M'. On the other
hand there are a point N of Σ and neighborhoods $V(M_o)$,
$V(M_o')$, $V(N)$ of M_o, M_o', N respectively in Σ such that the
planes of support at all points of $V(N)$ do not intersect
$V(M_o)$ nor $V(M_o')$ (it is here that the assumption that V
is not a double cone is used); an elementary geometrical argument then gives an upper bound < 1 for s. Historically,
such an argument would have been barely possible in the late
1870's, but I strongly doubt that C. Neumann was familiar
enough with the use of the "Bolzano-Weierstrass" theorem (as
it was called at that time) to have thought of it. He was

apparently satisfied with the fact that for simple convex sets,
such as ellipsoids, it was possible to compute explicity an
upper bound < 1 for s.

C. Neumann dealt in the same way with the Dirichlet problem
in the plane, with a similar gap in his proof.

For a long time, the restrictions on the surface Σ in all
the existence proofs of the Dirichlet problem were thought to
be imperfections of the methods of proof; but in 1912, Lebesgue gave an example (in 3 dimensions) of a bounded open set V
(homeomorphic to a ball) such that there is a continuous
function on the boundary Σ of V, for which the Dirichlet
problem has no solution ([138], vol. IV, p.131). This was
the starting point of modern Potential theory, where, on one
hand, the initial formulation of the Dirichlet problem is modified in such a way that it always has a unique "solution"
for any bounded domain, the word "solution" being interpreted
in some "weak" sense; on the other hand, the behavior of
these "weak" solutions on the boundary of the domain is investigated under various conditions [30]. The detailed
history of that extensive theory is outside the scope of this
book.

CHAPTER III

THE EQUATION OF VIBRATING MEMBRANES

§1 - H.A. Schwarz's 1885 paper

The same physical arguments which lead to the equation of vibrating strings (Chap.I, §2, equation (7)) apply to the small vibrations of a membrane which at rest is in the plane Oxy, and has a constant density: if $z = u(x,y,t)$ is the equation of its surface at time t, the function u satisfies the equation

$$(1) \qquad \frac{\partial^2 u}{\partial x^2} + \frac{\partial^2 u}{\partial y^2} = \frac{\partial^2 u}{\partial t^2}$$

(for suitable units of length and time). The usual method of "separation of variables" consists here in looking for solutions $u(x,y,t) = v(x,y)w(t)$ and one finds for v the equation (also called "Helmholtz's equation")

$$(2) \qquad \frac{\partial^2 v}{\partial x^2} + \frac{\partial^2 v}{\partial y^2} + \lambda v = 0$$

for a constant λ. If in addition the membrane at rest is a bounded portion Ω of the plane and is <u>fixed</u> at its boundary Σ (which means that $u(x,y,t) = 0$ for all t if $(x,y) \in \Sigma$), λ must be > 0, $w(t) = \sin \sqrt{\lambda}\, t$, and one has to find a solution v of (2) which vanishes on Σ and is not identically 0. Contrasting with the easy solution of the correspond-

47

ing problem for the vibrating string, the elucidation of that problem was going to challenge the ingenuity of mathematicians during the whole second half of the XIXth century.

Experimental evidence, as well as the explicit solution of the problem for very special domains Ω, such as a rectangle or a disk, showed that, just as in the case of the vibrating string, solutions of (2) vanishing on Σ and not identically 0 could only exist when λ was equal to one of an infinite sequence (λ_n) of real numbers > 0 (the "eigenvalues" of the problem), tending to $+\infty$.

The first attempt to prove such a result for general domains Ω was made by H. Weber in 1869 [224], by an adaptation of the variational method used by Riemann for the Dirichlet problem. Using Green's formula (Chapter II, formula (15)) he first shows that if μ_1, μ_2 are two distinct eigenvalues, v_1, v_2 corresponding "eigenfunctions", then

$$(3) \qquad (\mu_1 - \mu_2) \iint_\Omega v_1(x,y) v_2(x,y) dxdy = 0$$

from which he deduces, as Poisson had done for ordinary differential equations (Chap.I, §3), that the eigenvalues are necessarily real numbers. To determine the smallest eigenvalue λ_1, he considers the Dirichlet integral

$$(4) \qquad F(v) = \iint_\Omega \left(\left(\frac{\partial v}{\partial x}\right)^2 + \left(\frac{\partial v}{\partial y}\right)^2 \right) dxdy$$

for C^2 functions v in $\bar{\Omega}$, equal to 0 on Σ and subject to the additional constraint

$$(5) \qquad \iint_\Omega v^2 \, dxdy = 1.$$

He assumes, as Riemann, that in this set $\mathfrak{F}_1$ of functions, there is one for which $F(v)$ is equal to its greatest lower bound λ_1, and by the usual methods of the Calculus of variations, he shows that this function v_1 is a solution of (2) for $\lambda = \lambda_1$.

He next considers the subset $\mathfrak{F}_2$ of $\mathfrak{F}_1$ defined by the additional condition

$$(6) \qquad \iint_{\Omega} v(x,y)v_1(x,y)dxdy = 0,$$

takes the function $v_2 \in \mathfrak{F}_2$ for which $F(v_2)$ is equal to its greatest lower bound λ_2, and shows that v_2 is a solution of (2) for $\lambda = \lambda_2$. The induction process is then obvious, and Weber concludes that he has proved the existence of an increasing infinite sequence (λ_n) of positive eigenvalues to each of which there corresponds an eigenfunction v_n normalized by condition (5), and orthogonal to each other. But he does not try to prove that $\lim\limits_{n \to \infty} \lambda_n = +\infty$, nor that functions in $\mathfrak{F}_1$ possess a "Fourier expansion" $\sum\limits_{n} c_n v_n$ defined in the same manner as in the Sturm-Liouville problem (Chap.I, §3, formula (33)) (a result which he states however, without proof).

Weber's proofs were of course subject to the general criticisms of Weierstrass against the Calculus of variations, but no one seems to have tried to find more rigorous ones until 1885. In that year, H.A. Schwarz published a long paper on the theory of minimal surfaces, in which he had to consider a type of equation slightly more general than (2):

(7)
$$\frac{\partial^2 v}{\partial x^2} + \frac{\partial^2 v}{\partial y^2} + \lambda^2 \, pv = 0$$

where   p   is a continuous function in a domain   D,   with values > 0;   his arguments apply for any such function, but in
fact he is only interested in the particular case $p(x,y) =$
$= 8/(1+x^2+y^2)^2$ [195, vol.I, p.223-269].

Schwarz's paper is extremely remarkable by the originality
of its methods, which do not seem to have been inspired by any
previous work; it may be that the study of the Sturm-Liouville
problem led him to arguments which later could be transferred
almost _verbatim_ to general integral equations with symmetric
kernels (see Chap.V, §2), but there is no hint in his paper
of such an influence, and in fact he quotes nobody, not even
Weber.

His starting point is not the problem of existence of eigenvalues λ^2 for equation (7), but a "Dirichlet problem" for
the equation

(8) $\Delta w + \xi pw = 0$

depending on a parameter ξ;   he limits himself to the case
where   w   is subject to the condition of being equal to 1 on
the boundary Γ of D. Using the time honored method of representing the solution as a power series in ξ (Chap.II,§1)

(9) $w = w_0 + \xi w_1 +\ldots+ \xi^n w_n +\ldots$

he takes for w_0 the constant function equal to 1, and imposes on the w_n for $n \geq 1$ to vanish on Γ; they are then
determined inductively by the equations

(10) $\qquad \Delta w_n + p w_{n-1} = 0 \quad$ for $\quad n \geq 1.$

He assumes that the Green function $G(M,P)$ for the domain D exists (remember that he himself had proved that existence in extensive cases (Chap.II, §4)); the properties of that function implied that for any function f continuous in $\bar{D}$ the equation

(11) $\qquad\qquad\qquad\qquad \Delta w + f = 0$

has a unique solution vanishing on Γ, given by the formula

(12) $\qquad w(M) = \dfrac{1}{2\pi} \iint\limits_D f(P)G(M,P)d\omega \qquad$ (with $d\omega = dxdy$).

Therefore his functions w_n are given explicitly by

(13) $\qquad\qquad w_n(M) = \dfrac{1}{2\pi} \iint\limits_D p(P)w_{n-1}(P)G(M,P)d\omega.$

One must now investigate the convergence of (9) for small enough values of $|\xi|$, and it is here that Schwarz's original contributions begin. His main tool is the inequality named after him[*]

$$\left(\iint\limits_D fg\; d\omega \right)^2 \leq \left(\iint\limits_D f^2 d\omega \right)\left(\iint\limits_D g^2 d\omega \right)$$

for any two functions f, g continuous in $\bar{D}$; this gives from (13)

(14) $\quad 4\pi^2 (w_n(M))^2 \leq \left(\iint\limits_D p^2(P)G^2(M,P)d\omega \right)\left(\iint\limits_D w_{n-1}^2(P)d\omega \right)$

$$\leq A\left(\iint\limits_D w_{n-1}^2(P)d\omega \right)$$

---

[*] That inequality had been discovered by Buniakowsky in 1859, but does not seem to have been noticed nor used by many mathematicians before 1885. It is of course a direct generalization of the corresponding inequality for finite sums, which goes back at least to Cauchy.

where A is a constant independent of n (due to the properties of the Green function of a bounded domain). Schwarz
is thus led to study the numbers

$$(15) \qquad W_{n,k} = \iint_D pw_k w_{n-k} d\omega$$

which, using the <u>symmetry</u> of the Green function, he shows are
independent of k, so that $W_{n,k} = W_{n,0}$, which he writes W_n.
He also proves that

$$(16) \qquad W_{n,k} = \iint_D \left(\frac{\partial w_{k+1}}{\partial x} \frac{\partial w_{n-k}}{\partial x} + \frac{\partial w_{k+1}}{\partial y} \frac{\partial w_{n-k}}{\partial y} \right) dxdy.$$

Finally, using the Schwarz inequality, he obtains the relation

$$(17) \qquad W_n^2 \le W_{n-1} W_{n+1},$$

hence the sequence of numbers W_n/W_{n-1} is increasing; on the
other hand, integrating (14) gives $W_{2n} \le BW_{2n-2}$ for a constant B independent of n, and therefore the limit of the
sequence (W_n/W_{n-1}) is a finite number c > 0. It follows
then from (14) that the series (9) is absolutely and uniform-
ly convergent in $\bar{D}$ for $|\xi| < 1/\sqrt{c}$; the properties of the
Green function enable one to show that the derivatives of w
are also given by convergent series obtained by differentiating (9) termwise, and that w then satisfies (8) and is equal
to 1 on the boundary Γ.

 But Schwarz goes one step further. He proves that when
ξ = $1/\sqrt{c}$, the general term of the series (9) tends uniformly
to a limit U_1 which is not identically 0 in $\bar{D}$ but vanishes on the boundary and is solution of

$$(18) \qquad \Delta w + (1/c)pw = 0.$$

He has thus proved the existence of the <u>smallest eigenvalue</u>
$\lambda_1^2 = 1/c$ of the equation (8) for functions vanishing on the
boundary, and of the corresponding eigenfunction.

It should be observed here that these developments in fact
are just another treatment of a "crypto-integral" equation
(which Schwarz does not write, however). If one writes
$w = w_0 + \xi v$ and "solves" equation (8) by formula (12) (using
the same idea as Liouville in 1837 to obtain his "Volterra integral equation" (Chap.II, §1 and Chap.I, §3, equation (35))),
one gets for v this time a "Fredholm integral equation"

$$(19) \qquad v(M) = g(M) + \frac{\xi}{2\pi} \iint_D p(P)G(M,P)v(P)d\omega$$

with
$$g(M) = \frac{1}{2\pi} \iint_D G(M,P)p(P)d\omega.$$

Schwarz's procedure is therefore essentially the same as
C. Neumann's for the Dirichlet problem (Chap.II, §4), at least
as a starting point; the main difference is in the emphasis
put by Schwarz on the dependence on the parameter ξ.

To appreciate the originality and power of Schwarz's method,
it is perhaps not superfluous to show how it can be translated, almost without change, in the theory of self-adjoint compact operators in a separable Hilbert space E. Suppose U
is such an operator in E, which in addition we suppose
<u>positive</u>, i.e. $(U \cdot f | f) \geq 0$ for all $f \in E$. The spectrum of
U then consists in a decreasing sequence (finite or infini-
te) $\mu_1 \geq \mu_2 \geq \ldots \geq \mu_n \geq \ldots > 0$, where each μ_n is an eigenvalue counted a number of times equal to its multiplicity;
0 is always in the spectrum but $\text{Ker}(U)$ may be reduced to 0
or have infinite dimension; for each μ_n there is an eigen-

vector φ_n of norm 1, such that E is the <u>Hilbert sum</u> of
the one-dimensional spaces $\mathbb{C}\varphi_n$ and of $\text{Ker}(U)$. Let

$$w_o = \sum_n d_n \varphi_n + w_o'$$

with $w_o' \in \text{Ker}(U)$, be the expression of a vector $w_o \in E$
for that decomposition. Then, for any $m \geq 1$, we have

$$U^m \cdot w_o = \sum_n \mu_n^m d_n \varphi_n$$

and therefore the Schwarz series (9) is equal to

$$w = \sum_{m=0}^{\infty} \xi^m U^m \cdot w_o = \sum_n \left(\sum_{m=0}^{\infty} \xi^m \mu_n^m \right) d_n \varphi_n = \sum_n \frac{d_n}{1 - \xi \mu_n} \varphi_n$$

provided $|\xi| < \mu_1^{-1}$, and we have $w = \xi U \cdot w + w_o - w_o'$. For
$\xi = 1/\mu_1$,

$$\xi^m U^m \cdot w_o = \sum_n \xi^m \mu_n^m d_n \varphi_n = \sum_n \left(\mu_n / \mu_1 \right)^m d_n \varphi_n$$

tends to $d_1 \varphi_1$ if μ_1 is a simple eigenvalue, to the sum of
the $d_n \varphi_n$ such that $\mu_n = \mu_1$ in general. Finally, we have

$$W_m = (U^m \cdot w_o | w_o) = (U^{m-k} \cdot w_o | U^k \cdot w_o) = \sum_n \mu_n^m d_n^2$$

from which it follows that if the d_n such that $\mu_n = \mu_1$ are
not all 0, the ratio W_m / W_{m-1} tends to μ_1; furthermore,
one has

$$W_{2m} = \| U^m \cdot w_o \|^2 = | (U^{m+1} \cdot w_o | U^{m-1} \cdot w_o) | \leq \| U^{m+1} \cdot w_o \| \cdot \| U^{m-1} \cdot w_o \|$$

$$= \left(W_{2m-2} W_{2m+2} \right)^{\frac{1}{2}} .$$

To get the inequality $W_m \leq \left(W_{m-1} W_{m+1} \right)^{\frac{1}{2}}$ for all integers m,
it is enough to consider the unique compact positive operator
V such that $V^2 = U$, and apply to V the preceding argument.
Of course the concept of the "square root" of a positive self-

adjoint operator was not available to Schwarz, and this is why
he had to use the expression (16) for his numbers W_n.

In 1893, E. Picard published a short Comptes-Rendus Note
([172], vol.II, p.545-550) in which he went one step further.
For any point $M \in D$, the function $w(M,\xi)$ given by Schwarz's
series (9) is holomorphic in the circle $|\xi| < \xi_1 = 1/\sqrt{c}$, and
Picard investigates the <u>analytic continuation</u> of $\xi \mapsto w(M,\xi)$
beyond that circle: he shows that such a continuation exists
in a circle $|\xi| < \xi_2$ of radius independent of $M \in D$, and
that it has a <u>simple pole</u> with residue $-\xi_1 U_1(M)$ at the point
ξ_1. He limits himself to the case in which $p = 1$ and Γ is
convex and smooth, and his idea is to adapt the method of
C. Neumann (Chap.II, §4) to evaluate the differences
$|\xi_1^n w_n - \xi_1^{n-1} w_{n-1}|$; with apparently the same gap as in
Neumann's argument (the details are not given in the Note) he
"proves" that there are constants C and $q < 1$ independent
of M, such that $|\xi_1^n w_n - \xi_1^{n-1} w_{n-1}| \leq Cq^n$, hence $|\xi_1^n w_n - U_1| \leq$
$\leq C'q^n$ for another constant C', hence his result. Writing

$$w = \frac{U_1}{1-(\xi/\xi_1)} + v$$

he looks for a power series development

(20) $v = v_0 + \xi v_1 + \ldots + \xi^n v_n + \ldots$

similar to (9) but which should converge in a circle $|\xi| < \xi_2$
with $\xi_2 > \xi_1$. He determines the v_n by the successive approximations

$$\Delta v_0 - \xi_1 U_1 = 0, \qquad \Delta v_n + v_{n-1} = 0 \text{ for } n \geq 1$$

with the boundary conditions: $v_0 = 1$ on Γ and $v_n = 0$ on

Γ for $n \geq 1$. Introducing numbers similar to the $W_{n,k}$ of

Schwarz, he is able to prove that the radius of convergence

ξ_2 of (20) is finite, but he cannot show that there is an

eigenfunction corresponding to ξ_2 and vanishing on Γ.

§2 - The contributions of Poincaré

In 1890, H. Poincaré published in the American Journal of

Mathematics a long paper developing some of his research done

since 1887, which had been announced in three Comptes-Rendus

Notes ([177], vol.IX, p.15-113). The paper consists of two

completely independent parts; in the first, he describes in

detail his "sweeping-out" method for the solution of the Di-

richlet problem (Chap. II, §4). The second part is devoted

to the cooling off problem in the theory of heat, which had

been treated by Fourier in some particular cases, for instance

the cooling off of a sphere when the temperature is a function

of the distance to the center (Chap.I, §2). The general

cooling off problem had been presented by Fourier in the fol-

lowing form: given a solid body V of constant density, iso-

tropic for the propagation of radiations, one has to find the

temperature $u(x,y,z,t)$ inside V, as a function of the

coordinates x, y, z and the time t, when the outside tem-

perature is 0. Fourier shows that the function u must sa-

tisfy inside V an equation (where a is constant)

(21)
$$\frac{\partial u}{\partial t} = a^2 \Delta u$$

and in addition is subject to the boundary condition on the

surface Σ of V

(22)
$$\frac{\partial u}{\partial n} + hu = 0$$

where $\frac{\partial u}{\partial n}$ is the normal derivative (towards the exterior)
and h is a constant ≥ 0 (see Chap.IV, §4). The usual
method of "separation of variables" led to solutions of the
form $u(x,y,z,t) = e^{-\lambda a^2 t} v(x,y,z)$, where v should be a
solution of the Helmholtz equation

(23)
$$\Delta v + \lambda v = 0$$

with a <u>different</u> boundary condition from the one deriving
from the equation of vibrating membranes, namely

(24)
$$\frac{\partial v}{\partial n} + hv = 0 \quad \text{on} \quad \Sigma.$$

In his 1869 paper, H. Weber had also considered that problem,
but he had only described his variational method to obtain
eigenvalues and eigenfunctions for the particular case h=0.
Poincaré apparently was unaware of Weber's paper and never
mentioned it in his own work; what he does in 1890 is first
to repeat Weber's arguments for the general boundary condition (24), replacing the Dirichlet integral by the function

(25)
$$F(v) = h \iint_{\Sigma} v^2 d\sigma + \iiint_{V} \left(\left(\frac{\partial v}{\partial x}\right)^2 + \left(\frac{\partial v}{\partial y}\right)^2 + \left(\frac{\partial v}{\partial z}\right)^2 \right) d\omega.$$

Having thus obtained an increasing infinite sequence (λ_n) of
eigenvalues and the corresponding sequence (v_n) of eigenfunctions, Poincaré is of course aware of the non rigorous
character of his "proof"; however, having for the time being
no better arguments at his disposal, he takes for granted the
existence of λ_n and v_n, and proceeds to study them in more

detail, and in the first place to prove that the sequence (λ_n)
tends to $+\infty$, a question which Weber had not been able to
answer. In his attack on that problem, it is quite remarkable
to see Poincaré introducing a whole batch of completely new
ideas. In the first place, he considers the eigenvalues as
functions $\lambda_n(h,V)$ of the constant h and the domain V and
begins to study the way in which they depend on h and V,
a trend of thought which will later blossom in the work of
H. Weyl and R. Courant, and even now has not entirely lost its
interest. Poincaré first shows that, for V fixed, $\lambda_n(h,V)$
is increasing with h, by an application of Green's formula
to the eigenfunctions $v_n(h,V)$, $v_n(h',V)$ corresponding to
two values of h; as he wants to prove that $\lambda_n(h,V)$ tends
to $+\infty$, he can assume that $h = 0$, which implies that
$\lambda_1(0,V) = 0$ and $v_1(0,V)$ is a constant.

 The second idea is to decompose V into a union of smaller
solids $V_1,V_2,\ldots,V_p$; the variational definition of λ_n
enables him to prove that if $p \le n-1$, $\lambda_n(0,V)$ is at least
equal to the smallest of the numbers $\lambda_2(0,V_1),\ldots,\lambda_2(0,V_p)$.
Poincaré is thus led to minorize $\lambda_2(0,V)$ by a number depending only on the geometry of V; by definition (since $h = 0$),
this means finding a lower bound of the expression

$$(26) \qquad \frac{\iiint_V \left(\left(\frac{\partial v}{\partial x}\right)^2 + \left(\frac{\partial v}{\partial y}\right)^2 + \left(\frac{\partial v}{\partial z}\right)^2 \right) d\omega}{\iiint_V v^2 d\omega}$$

where v is a C^2 function in $\bar{V}$, subject to the condition

$$(27) \qquad \iiint_V v \, d\omega = 0.$$

He assumes V is <u>convex</u>; using polar coordinates and the
standard methods of the Calculus of variations, he obtains as
lower bound

(28) $C \cdot vol(V)/(diam(V))^5$

where C is an absolute constant; one should here stress the
fact that this Poincaré inequality is the first example of
what we now call "<u>a priori</u>" <u>inequalities</u> (cf. Chap. IX, §4).
Returning to the minoration of $\lambda_n(0,V)$, he takes $p = n-1$,
assumes that V can be decomposed in n-1 solids V_j which
are convex and have a diameter tending to 0 with 1/n and
such that the ratio of their volume to the fifth power of
their diameter tends to +∞ with n; this gives him his
conclusion.

Poincaré's next step is to investigate how the knowledge of
the λ_n and v_n gives the solution of the cooling off problem, when the temperature $u(x,y,z,0)$ is a known function
$f(x,y,z)$ in V at time t = 0. Fourier's method consists
in writing

(29) $u(x,y,z,t) = \sum_{n=1}^{\infty} c_n \exp(-\lambda_n a^2 t) v_n(x,y,z)$

which gives for the unknown coefficients c_n the condition
that $f = \sum_{n=1}^{\infty} c_n v_n$, hence, from the orthogonality relations,
$c_n = \iiint_V f v_n \, d\omega$. But Poincaré, no more than Weber, is not at
that time able to prove that this Fourier expansion converges
to the function f in V. However, taking his cue from
Tchebychef's results in approximation theory, he shows (by a
clever use of Green's formula) that the integral

$$S_n = \iiint_V (u - \sum_{k=1}^{n} c_k \exp(-\lambda_k a^2 t)v_k)^2 \, d\omega$$

satisfies an inequality $S_n \leq C \cdot \exp(-\lambda_{n+1} a^2 t)$ where C is independent of n and t; in other words, for $t > 0$, he proves the convergence of the series in (29) in what we now call the topology of Hilbert space[(*)].

The final section of Poincaré's paper (if we except a kind of postscript which we will discuss later in Chapter IV, §4) is devoted to the general study of the eigenfunctions v_n

---

[(*)] The method of least squares of Legendre-Gauss had led Tchebychef to define a "best approximation" to a function F, by a linear combination $\sum_{j=1}^{N} a_j \psi_j$ of given functions ψ_j ($1 \leq j \leq N$), by the condition that

$$\sum_{k=1}^{n} \rho(x_k)(F(x_k) - \sum_{j=1}^{N} a_j \psi_j(x_k))^2$$

be minimum, for given points x_k ($1 \leq k \leq n$) and given "weight" ρ. Gram, in 1883, generalized the problem by considering instead of a finite sum, an integral

$$(+) \qquad \int_a^b \rho(x)(F(x) - \sum_{j=1}^{N} a_j \psi_j(x))^2 \, dx$$

and he solved the problem in an original way, by applying to the ψ_j the "orthogonalization process" usually attributed to E. Schmidt [89]. He was thus reduced to the case in which the ψ_j form an orthonormal system (for the measure ρdx), where he showed that the a_j giving the best approximation are the "Fourier coefficients" $\int_a^b \rho(x)F(x)\psi_j(x)dx$. He went on to consider an _infinite_ orthonormal system (ψ_n) and investigated under which conditions the minimum value μ_n of the integral (+) tends to 0 when n increases to $+\infty$; he was able to see that this was linked to the "completeness" of the system (ψ_n), i.e. the fact that no function other than the constant 0 is orthogonal to all ψ_n. It is unlikely that Poincaré had any knowledge of Gram's paper.

(their existence being admitted). In general, if v satisfies (23) and (24), use of Green's formula shows that there
is a formula similar to Green's expression of the potential
(Chap.II, §3, formula (17))

$$(30) \qquad\qquad -4\pi v(M) = \iint\limits_{\Sigma} v\left(\frac{\partial T}{\partial n} + hT\right)d\sigma$$

where T (replacing the function 1/r). is now $\exp(i\sqrt{\lambda}\ r)/r$.
Using that formula, he is able to show, after a rather long
discussion (patterned on the study of double layer potentials
but more difficult), that v is continuous in $\bar{V}$, and to
obtain bounds for its derivatives in V.

The second paper devoted by Poincaré to the equation of vibrating membranes ([177], vol.IX, p.123-196) is even more
original. It is likely that in 1890, he was not aware of
Schwarz's paper of 1885. The publication of Picard's note in
1893 immediately attracted his attention, and in a few months
he had seen that by combining Schwarz's method and his "a
priori" inequality of 1890, he could go beyond Picard and
prove the analytic continuation of the function $\xi \mapsto w(M,\xi)$
as a meromorphic function in the whole complex plane, obtaining at the same time the existence of the long sought eigenvalues and eigenfunctions for the Helmholtz equation (with
the same boundary condition as Schwarz).

Poincaré starts with a simplification and an improvement of
his inequality for the expression (26); using Schwarz's inequality, he is able to replace his lower bound (28) by
$C/(\text{diam}(V))^2$ for a convex solid V. He then only assumes
that for a general solid V it is possible to decompose it

in convex solids having arbitrary small diameters, and uses
this idea of decomposition to prove the following crucial
lemma: given p arbitrary C^2 functions $F_1, F_2, \ldots, F_p$ in
$\bar{V}$, it is possible to choose p numbers $\alpha_1, \ldots, \alpha_p$ in such
a way that, for $v = \alpha_1 F_1 + \ldots + \alpha_p F_p$, one has $\iiint_V v \, d\omega = 0$
and the ratio (26) is <u>at least</u> L_p, where L_p is a number
which only depends on V and p (and not on the F_j) and
<u>tends to</u> $+\infty$ with p. This is simply done by decomposing V
in the union of p-1 convex subsets V_j, and choosing the
coefficients α_j by the p-1 conditions $\iiint_{V_j} v \, d\omega = 0$
$(1 \leq j \leq p-1)$.

Poincaré, as Picard, limits himself to the case in which the
function p in equation (8) is the constant 1, but considers
a problem which slightly generalizes Schwarz's, namely he
looks for a function v solution of

(31) $\Delta v + \xi v + f = 0$

and vanishing on the boundary Σ, with f an <u>arbitrary</u> C^∞
function (if in Schwarz's equation (8) with p = 1, one
writes $w = w_0 + \xi v$, the equation for v is (31) with $f = w_0$);
he will make a very clever use of this arbitrariness. He
starts by observing that Schwarz's method works just as well
for arbitrary f as for f = 1, and proves the existence of
the solution of (31) vanishing on Σ for <u>small enough</u> $|\xi|$;
he writes it

$$v = [f, \xi] = v_0 + \xi v_1 + \ldots + \xi^n v_n + \ldots$$

with $\Delta v_0 + f = 0$, $\Delta v_n + v_{n-1} = 0$ for $n \geq 1$, the v_n all
vanishing on Σ.

For any given integer p, he introduces p arbitrary coefficients $\alpha_1,\ldots,\alpha_p$ and forms the function (defined at least for small ξ)

$$(32)\quad w = [\alpha_1 f + \alpha_2 v_0 + \cdots + \alpha_p v_{p-2}, \xi] = w_0 + w_1\xi + \cdots + w_n\xi^n + \cdots$$

Next, applying his lemma for the evaluation of the Schwarz integrals W_n _corresponding to_ w, he is able to show that, for a suitable _choice_ of the α_j, the series (32) _converges for_ $|\xi| \le L_p$ (uniformly in $\bar V$). But if one writes $u_j = [v_{j-2}, \xi]$, one has

$$(33)\quad \begin{cases} \alpha_1 v + \alpha_2 u_2 + \cdots + \alpha_p u_p = w \\ v - u_2\xi = v_0 \\ u_2 - u_3\xi = v_1 \\ \cdots\cdots\cdots\cdots \\ u_{p-1} - u_p\xi = v_{p-2} \end{cases}$$

a linear system from which Cramer's formulas give

$$(34)\qquad v = P/D$$

with

$$D = \begin{vmatrix} \alpha_1 & \alpha_2 & \alpha_3 & \cdots & \alpha_{p-1} & \alpha_p \\ 1 & -\xi & 0 & \cdots & 0 & 0 \\ 0 & 1 & -\xi & \cdots & 0 & 0 \\ \cdots & \cdots & \cdots & \cdots & \cdots & \cdots \\ 0 & 0 & 0 & & 1 & -\xi \end{vmatrix} = \alpha_p - \alpha_{p-1}\xi + \alpha_{p-2}\xi^2 - \cdots + (-1)^{p-1}\alpha_1\xi^p$$

and

$$
P = \begin{vmatrix}
w & \alpha_2 & \alpha_3 & \cdots & \alpha_{p-1} & \alpha_p \\
v_o & -\xi & 0 & \cdots & 0 & 0 \\
v_1 & 1 & -\xi & \cdots & 0 & 0 \\
\cdot & \cdot & \cdot & \cdot \cdot \cdot & \cdot & \cdot \\
v_{p-2} & 0 & 0 & \cdots & 1 & -\xi
\end{vmatrix}
$$

which shows that P, as w, is equal to a series

(35) $P = P_o + P_1\xi + \ldots + P_n\xi^n + \ldots$

where the P_n are C^∞ functions in $\bar{V}$ vanishing on Σ, the
series being uniformly convergent in $\bar{V}$ for $|\xi| < L_p$, and
all derivatives of P (with respect to ξ or to x,y,z)
being obtained by derivating termwise the series. This shows
that $\xi \mapsto v(M,\xi)$ extends to a <u>meromorphic</u> function in $|\xi| < L_p$
with a finite number of poles, which are roots of $D(\xi) = 0$
and <u>independent</u> of M; it is easy to show, using Green's
formula, that these poles are all <u>simple</u>. As this is true
for any p, $\xi \mapsto v(M,\xi)$ extends to a meromorphic function in
the <u>whole complex plane</u>, with simple real and positive poles
independent of M; furthermore, for each one of these poles
λ_n, the function $P(M,\lambda_n)$ satisfies $\Delta P + \lambda_n P = 0$; in
other words, one has found for each λ_n an eigenfunction u_n
corresponding to that eigenvalue. In addition, Poincaré's
<u>a priori</u> inequality enables him to show that $\lambda_n \geq c \cdot n^{2/3}$,
where c is a constant.

The remainder of Poincaré's 1894 paper is devoted to two
questions:

 A) In the last 4 sections of the paper, he takes up again
the problem of Fourier expansions (when the boundary condition

is $v=0$). Attaching to the function f its "Fourier coefficients" $c_n = \iiint_V f u_n \, d\omega$ (where the eigenfunctions u_n have been normalized by $\iiint_V u_n^2 \, d\omega = 1$), he first deduces from the relations $\Delta u_n + \lambda_n u_n = 0$ and Schwarz's inequality, that $|u_n| \leq A\lambda_n$ in $\bar{V}$ (A constant), and that the c_n are uniformly bounded. From that it follows that for ξ different from the eigenvalues the unique solution of (31) vanishing on Σ is given by the absolutely and uniformly convergent series

$$(36) \qquad v = - \sum_n \frac{c_n u_n \xi^2}{\lambda_n^2 (\xi - \lambda_n)} + v_o + v_1 \xi;$$

in addition, Poincaré shows that if the series $\sum_n c_n u_n$ is absolutely convergent, its sum is equal to f; he cannot prove that for "arbitrary" functions f (probably at least c^4), vanishing on Γ, the series converges, but he proves absolute convergence when in addition Δf and $\Delta^2 f$ also vanish on Σ.

B) Before returning to the question of Fourier expansions, Poincaré had tried to extend his results on the existence of eigenvalues and eigenfunctions for the boundary condition (24) of the cooling off problem. He realizes that Schwarz's method would work, and therefore also his own existence theorem, <u>provided</u> one could prove the existence of a "Green function" for the Laplace equation with that new boundary condition, i.e. a function $G(M,P)$ having the same properties as the usual Green function, with the exception that, for $M \in V$, $P \mapsto G(M,P)$ satisfies (24) on the boundary.

<u>In the special case</u> $h=0$, C. Neumann, in his work on the

Dirichlet problem, had shown how to obtain such a "Green
function" (also named "Neumann function") when V is convex
and not a double cone. He had observed that by changing the
sign before the integral in the Beer-Neumann equation (Chap.
II, §4, formula (23)), the solution of that new equation gave
a density ρ such that the corresponding double layer potential, in the <u>exterior</u> of V (complement of $\bar{V}$) is harmonic,
tends to O at infinity and to -g on the boundary Σ (a
solution to what is called the "exterior Dirichlet problem").
From this result, he had shown how to obtain a solution of
what is now called the <u>Neumann problem</u> for the Laplace equation: find in V a harmonic function u   such that   u   is
continuous in $\bar{V}$ and has on Σ a normal derivative $\frac{\partial n}{\partial u}$
equal to a <u>given</u> continuous function g; a necessary condi-
tion for the existence of the solution (deduced from Green's
formula applied to u and the constant 1) is that $\iint_{\Sigma} g \, d\sigma = 0$.
Neumann proves that this condition is sufficient (the solu-
tions being determined up to an additive constant): he considers the <u>simple layer</u> potential w defined by the density
$\frac{1}{4\pi}$ g; it is continuous on Σ and its normal derivative
jumps by -g when crossing Σ from the interior to the exterior. Neumann next takes the <u>double layer</u> potential v,
solution of the <u>exterior</u> Dirichlet problem which tends to -w
on Σ. Then the function u = v+w is harmonic outside Σ,
and O in the exterior of V; as the normal derivative of
v is the same on both sides of Σ, it follows at once that
$\frac{\partial u}{\partial n}$ tends to g from Σ from the interior of V, and therefore solves the Neumann problem.

C. Neumann had not been able to solve the corresponding prob-
lem when the boundary condition is $\frac{\partial u}{\partial n} + hu = g$ for a
constant $h > 0$. Poincaré tried to solve the problem by representing u as a power series in h, $u = u_0 + hu_1 + \ldots +$
$+ h^n u_n + \ldots$, and was indeed able to obtain in that way
(using Neumann's results) a series convergent for all $h \geq 0$,
uniformly in $\bar{V}$; however, for the first derivatives, his
method could only prove uniform convergence in compact sub-
sets of V, so that it was impossible to give meaning to $\frac{\partial u}{\partial n}$
on the boundary Σ, and to show that u was indeed a solu-
tion of the problem. The most interesting result in this
attempt is that Poincaré, probably for the first time in history, arrives at the idea of "weak" solution of a boundary
problem; he shows that his function u is such that, for any
function v which is C^2 in $\bar{V}$, one has

$$(37) \quad \iiint_V u \Delta v \, d\omega + \iint_\Sigma gv d\sigma = \iint_\Sigma (\frac{\partial v}{\partial n} + hv) u d\sigma$$

and adds that "physically" this is equivalent to a genuine
solution.

The last of the three long papers of Poincaré on partial
differential equations was written in 1895 ([177], vol.IX,
p.202-272). Although it is the one which contains the smallest number of new results, it probably had a greater influence than the others. From his work both on the Dirichlet
problem and on the equation of vibrating membranes, Poincaré
had become convinced that there were also "eigenvalues" and
"eigenfunctions" linked to the Dirichlet problem. For us this
is completely obvious, for if we look for a solution of $\Delta u = 0$

taking given values on the boundary Σ of V, we extend the function g given on Σ to a C^2 function h in $\bar{V}$ (when this is possible); replacing u by v = u-h, we have to find a solution of $\Delta v + f = 0$, with $f = \Delta h$, which vanishes on Σ, and this is just the special case of Schwarz's problem for the equation (31) with $\xi = 0$.

At that time, however, nobody had yet thought of this simple argument[*], and Poincaré's reasoning is quite different and much more circuitous. He observes that one can formulate both the interior and exterior Dirichlet problems as special cases of the problem which consists in finding a double layer potential W (for a density on Σ) such that, for $s \in \Sigma$,

$$(38) \qquad W(s^-) - W(s^+) - \lambda(W(s^-) + W(s^+)) = 2\Phi(s)$$

where $W(s^-)$ is the limit of W at s along the interior normal, $W(s^+)$ its limit along the exterior normal, λ is a complex parameter and Φ a given function on Σ; the values $\lambda = 1$ and $\lambda = -1$ correspond respectively to the interior and the exterior Dirichlet problem. To this general problem Poincaré associates a new variational problem: for any simple layer potential Ψ defined by a density on Σ, he considers the ratio J/J', where J is the Dirichlet integral $\iiint (\text{grad } \Psi)^2 \, d\omega$ extended over V, and J' the integral of the same function, extended to the exterior of V. The usual non rigorous arguments lead him to conjecture: 1º the existence

---

[*]It is explicitly mentioned in 1909 by E.E. Levi [145, vol. II, p.302-313]; the first statement and proof of the existence of a continuous function in the whole space $\mathbb{R}^3$ extending a given function defined and continuous in a closed subset (i.e. what we now call the Tietze-Urysohn theorem) is due to Lebesgue in 1907 [138, vol.IV, p.99-100].

of an increasing sequence $0 = \lambda_0 \leq \lambda_1 \leq \ldots \leq \lambda_n \leq \ldots$ of

eigenvalues, and: $2^{\underline{o}}$ for each λ_i, the existence of a simple

layer potential Φ_i, such that, on Σ,

$$(39) \qquad \frac{\partial \Phi_i}{\partial n}(s^-) + \lambda_i \frac{\partial \Phi_i}{\partial n}(s^+) = 0;$$

in addition, for $i \neq j$, $\iiint_V \mathrm{grad}(\Phi_i) \cdot \mathrm{grad}(\Phi_j) d\omega = 0$. Nor-

malizing the Φ_i by $\iiint_V (\mathrm{grad}\ \Phi_i)^2\ d\omega = 1$, he assumes that

there is a Fourier expansion $\Phi = \sum_i c_i \Phi_i$ of the given func-

tion Φ on Σ, and "solves" the equation (38) by

$$W(s^-) = \sum_i A_i \Phi_i(s), \qquad W(s^+) = -\sum_i \lambda_i A_i \Phi_i(s)$$

with $A_i = 2c_i / (1 + \lambda_i - \lambda(1-\lambda_i))$.

All this is of course presented by Poincaré as purely con-

jectural, and as a motivation for his detailed study (by

methods inspired by those of Schwarz) of the ratio J/J',

which forms the central part of his 1896 paper; but the only

positive result he is able to deduce from his study is that

the Beer-Neumann series (Chap.II, §4) converges, not only for

convex domains V, but also for domains $V \subset \mathbb{R}^3$ having the

following property: when $\mathbb{R}^3$ is imbedded in the 3-dimensio-

nal sphere S_3 by adjoining a point at infinity, V can be

transformed into a ball by a homeomorphism of S_3 onto itself,

leaving fixed the point at infinity, and which is C^2 in S_3

as well as the inverse homeomorphism[(*)].

[(*)]Without the slightest justification, Poincaré claims as
"clear" the fact that this property holds for any bounded domain V such that the boundary Σ is a smooth simply connected
surface ([177],vol.IX,p.223-224). With the tools of modern Differential Topology, it is now possible to prove that theorem.
But in 1895, Poincaré was just beginning to formulate the first
notions of that theory, and one wonders if he realized the difficulties which lay in the way of a rigorous proof if he had
tried to write it down (when smoothness conditions on Σ and on
the homeomorphism are dropped, the result is known to be false,
a counterexample being the famous "Alexander horned sphere").

Almost immediately after the publication of Poincaré's papers,
several mathematicians were able to complete and extend his
results. In 1898, E. Le Roy [144] proved the existence of the
simple layer potentials Φ_i conjectured by Poincaré in his
1896 paper; he replaced the ratio J/J' by $(J+J')/I$, where
I is the surface integral $\iint_\Sigma \rho^2 \, d\sigma$, ρ being the density
on Σ corresponding to the simple layer potential Ψ, and
adapted the methods of Schwarz and Poincaré to the corresponding variational problem. In 1899, S. Zaremba [232], by a mo-
dification of the method of solution of Neumann's problem
used by Poincaré, could complete the latter's solution of the
"cooling off" problem, proving that the "weak" solution of
Poincaré was a genuine one. In 1901, Zaremba and W. Stekloff,
independently, finally showed that one could drop the global
topological property of the domain V which Poincaré and Le
Roy had used, and even weaken the "smoothness" conditions on
Σ; they made essential use of a paper of Liapounoff published 3 years earlier [147], in which he was able to prove the
existence of the normal derivative on Σ of the solution of
Dirichlet's problem under these less stringent conditions.

CHAPTER IV

THE IDEA OF INFINITE DIMENSION

§1 - Linear algebra in the XIXth century

I think that in order to understand the trend of ideas which led to Functional Analysis, it is useful to summarize the evolution of linear algebra during the XIXth century. Until around 1830, it had consisted in the study of systems of linear equations in any number of variables, with real or complex coefficients, most of the times limited to the case in which the number of equations was equal to the number of variables; the Cramer formulas gave the unique solution when the determinant of the system was not 0, but not much effort was spent on the elucidation of the other cases; the only result which was used occasionally was the fact that a system of m homogeneous equations in $n > m$ variables always had a non trivial solution (obvious by induction on m).

Linear changes of variables

$$(1) \qquad y_j = \sum_{k=1}^{n} a_{jk} x_k \qquad (1 \le j \le m)$$

had been familiar since the XVIIIth century (mostly for $m = n \le 3$). They naturally led to computations done, not on numbers, but on __rectangular arrays__ (a_{jk}) of numbers, which "represented" these changes of variables. Beginning with Gauss, this trend was systematized in the 1850's by Sylvester

71

and Cayley in the theory of <u>matrices</u>.

Ever since the invention of cartesian coordinates ("analytic geometry", as it came to be called in the XVIIIth century), mathematicians had known how to interpret geometrically computations on systems of 2 or 3 variables, and many had envisioned the possibility of similarly interpreting computations on systems of any number n of variables in a "geometry in n dimensions", which however would be devoid of "reality". After 1840, mainly under the influence of Hamilton and Cayley, this geometrical language was gradually adopted by more and more mathematicians, and had become commonplace at the end of the century. But in the XIXth century, after 1822 "geometry" essentially meant <u>projective</u> geometry, and most "geometric" interpretations of computations were done, not in the vector space $\mathbb{R}^n$ or $\mathbb{C}^n$, but in the complex projective spaces $\mathbb{P}_n(\mathbb{C})$; for instance, the relations (1) for $m = n$ were interpreted as defining also a <u>projective transformation</u> in $\mathbb{P}_{n-1}(\mathbb{C})$ sending the point of homogeneous coordinates (x_j) to the point of homogeneous coordinates (y_j), and the efforts of Grassmann and Peano to introduce vector spaces in an axiomatic way were persistently ignored until 1900.

Between 1850 and 1880 are proved the main theorems of linear algebra, concerning what are called the "reductions" of square matrices. One of these is the problem of finding, for a given square matrix U, an invertible matrix P such that PUP^{-1} has a "reduced" unique canonical form, which here (for complex matrices U) means a diagonal array of Jordan matrices; this is the way Jordan himself treats the problem, improving on a previous result of Grassmann, who had proved the existence of a "reduced" triangular matrix PUP^{-1} for any U (using already the intrinsic notion of endomorphism instead of the notion of square matrix).

Unfortunately, another type of "reduction" interfered with the preceding one. To a quadratic form $\sum_{j \leq k} a_{jk} x_j x_k$ corresponds the <u>symmetric</u> matrix $(a_{jk}) = U$, and it was well known since Cauchy that if U is <u>real</u> it is possible to find an invertible real matrix P (which may even be supposed to be <u>orthogonal</u>) such that PUP^{-1} would be a (real) <u>diagonal</u> matrix; this is equivalent to finding an orthogonal change of variables for which the quadratic form became equal to a linear combination $\sum_{j=1}^{n} \lambda_j y_j^2$ of squares, the λ_j being the elements of the diagonal matrix PUP^{-1}, or equivalently the roots (with their multiplicity) of the "characteristic equation" $\det(U - \lambda I) = 0$. Weierstrass, who was the first to find the "Jordan normal form" of a square complex matrix (which Jordan only discovered independently 2 years later[*]), presented it as a generalization of the "reduction" of a quadratic form, by considering a <u>bilinear form</u> $\sum_{j,k} a_{jk} x_j y_k$ (with $U = (a_{jk})$ an <u>arbitrary</u> square matrix) and applying to the x_j and y_j two "contragredient" changes of variables, i.e. such that the bilinear form $\sum_{j,k} x_j y_k$ remains invariant; this amounts to replacing U by a matrix PUP^{-1}. When, in 1878, Frobenius gave a systematic account of these results ([78], vol.I, p.343-405), he deliberately abandoned the lan-

---

[*] Jordan was not dealing with matrices having elements in $\mathbb{R}$ or $\mathbb{C}$, but with matrices having elements in a <u>finite field</u> ([123], p.114-126).

guage of matrices in favor of the language of bilinear forms, defining the "product" (<u>Faltung</u>) of two bilinear forms $A(x,y)$, $B(x,y)$, as $\sum\limits_{k=1}^{n} \dfrac{\partial A}{\partial y_k} \dfrac{\partial B}{\partial x_k}$! $(*)$

Finally, the concept of duality in <u>vector spaces</u> was completely foreign to mathematicians until 1900. Duality was well understood in the realm of <u>projective geometry</u> (it had been one of the big discoveries of the early XIXth century), as a bijection of points on planes (in projective space of 3 dimensions) and later as a bijection of points on hyperplanes in any number of dimensions. But linear forms were identified with the systems of their coefficients, "vectors" and "forms" being thus both "n-tuples" of numbers, which one had to distinguish, according to the way they behaved under changes of variables, by the awkward concepts of "contragredient" and "cogredient" systems. This identification of a vector space and its dual was reverberated in the identification of endomorphisms with bilinear forms, mentioned above$^{(**)}$.

To sum up, at the end of the XIXth century, the main results of linear and multilinear Algebra had been found but were expressed through insufficiently clarified notions. They could therefore be of no help to the generalizations of linear Algebra to infinite dimensional spaces which were called forth

---

$(*)$ In 1896, Pincherle reinterpreted Weierstrass's results in terms of endomorphisms ([173], vol. I, p.358-367).

$(**)$ In modern linear algebra, the space of endomorphisms of a finite dimensional vector space E is identified with the tensor product $E^*\otimes E$, whereas the space of bilinear forms on $E\times E$ is identified with $E^*\otimes E^*$.

by the development of Functional Analysis; these had to go
through the same painful stages, first linear equations, then
determinants, later bilinear forms, matrices, and only at the
very end vector spaces and linear maps; in other words, the
historical evolution, just as for finite dimensional linear
algebra, was exactly in the reverse order of what we now consider to be the logical order!

§2 - Infinite determinants

The first appearance of infinite systems of linear equations
in infinitely many unknowns seems to occur in Fourier's work
on the theory of heat. He has to determine an infinite sequence $(a_m)_{m \geq 1}$ of coefficients such that the relation

$$(2) \qquad\qquad 1 = \sum_{m=1}^{\infty} a_m \cos(2m-1)y$$

holds for all y ([67], vol.I, p.149). Fourier's idea is to
take derivatives of all orders of both sides of (2) and identify them for y = 0, which gives him the infinite system of
linear equations for the a_m

$$(3) \qquad \begin{cases} 1 = \sum_{m=1}^{\infty} a_m \\[2mm] 0 = \sum_{m=1}^{\infty} (2m-1)^2 a_m \\[2mm] 0 = \sum_{m=1}^{\infty} (2m-1)^4 a_m \\[2mm] \cdots\cdots\cdots\cdots \end{cases}$$

To solve it, he considers the first k equations where he
replaces the a_m for m > k by 0; he then solves that

system by Cramer's formulas, which give him a system of k

numbers $a_1^{(k)}, a_2^{(k)}, \ldots, a_k^{(k)}$,   and lets k   tend to infinity

in each expression of $a_m^{(k)}$ for fixed m. Using the formulas

giving Vandermonde determinants, he obtains

$$a_1^{(k)} = \frac{3^2 \cdot 5^2 \ldots (2k-1)^2}{8 \cdot 24 \ldots (4k^2 - 4k)}$$

tending to $a_1 = 4/\pi$,   and

$$\frac{a_{m+1}^{(k)}}{a_m^{(k)}} = \frac{2m-1}{2m+1} \cdot \frac{m+k}{m-k}$$

which gives him $a_m = (-1)^{m-1} \, 4/\pi (2m-1)$;   when later in his

book he proves the general formula giving the Fourier coef-

ficients, he can of course check that these values of the a_m

are correct. But he never bothered to give any justification

of his procedure, where all questions of convergence are com-

pletely disregarded; that procedure could of course be repeat-

ed for any infinite system

(4) $\sum_{k=1}^{\infty} a_{jk} x_k = b_j$ $(j = 1, 2, \ldots)$

but nobody undertook to justify it before 1885[*]. In that

year, P. Appell met such a system with $a_{jk} = a_k^j$ for a given

sequence (a_k), in a question relative to elliptic functions,

and used the same method as Fourier; his paper attracted

Poincaré's attention, and he showed that for such a "general-

ized Vandermonde system", the procedure was justified provided

---

[*]During that period, Fourier's method was used in two little known papers, one by Fürstenau in 1860 on the computation
of roots of an algebraic equation, and another by Kötteritzch
in 1870, for a system (4) in which the a_{jk} are 0 for j > k
(see [184], p.8-12).

the infinite product $F(z) = \prod_{k=1}^{\infty} (1-\dfrac{z}{a_k})$ was convergent for all complex numbers z.

The next year, he returned to the subject, in relation with a paper published in 1877 by the American astronomer and mathematician G.W. Hill on the lunar theory [114]. Hill proposed a new approach which rested on the integration of a second order differential equation

(5) $$w'' + (\sum_{n=-\infty}^{+\infty} \theta_n e^{nit})w = 0$$

where the θ_n are constants, and one looks for a solution of period 2π; Hill writes such a solution as a trigonometric series

(6) $$w = \sum_{n=-\infty}^{+\infty} b_n e^{i(n+c)t}$$

and substituting in (5), obtains for the coefficients b_n the infinite system of equations

(7) $$\sum_{k=-\infty}^{+\infty} \theta_{n-k}b_k - (n+c)^2 b_n = 0, \qquad -\infty < n < +\infty.$$

He probably was unaware of Fourier's procedure, but used a similar one, keeping this time the equations (7) for $-p \leq n \leq p$, replacing in these equations the b_m by 0 for $m < -p$ or $m > p$, and letting p tend to infinity in the solutions of the system thus obtained.

Poincaré considers a general system (4), where he supposes that $a_{jj} = 1$ for all j (one can always reduce (4) to such a system by dividing the j-th equation by a_{jj}, when $a_{jj} \neq 0$). His idea is to compare the determinant $D_n = \det(a_{jk})_{1 \leq j,k \leq n}$ to the product $P_n = \prod_{j=1}^{n} (\sum_{k=1}^{n} |a_{jk}|)$. It

is clear from the definition of a determinant that $|D_n| \leq P_n$,
and from the assumption on the diagonal terms, one has also
$|D_m - D_n| \leq P_m - P_n$; this inequality immediately gives Poincaré's
<u>sufficient</u> condition for the existence of $D = \lim_{n \to \infty} D_n$, name-
ly that the double sum $\sum_{j \neq k} |a_{jk}|$ be <u>finite</u>. Furthermore,
Poincaré shows that, when the k-th column of D is replaced
by a sequence (b_j) which is <u>bounded</u>, there is still convergence for the new "infinite determinant", and that there is a
unique bounded solution (x_k) of (4) given by the usual
Cramer formulas (with "infinite determinants" of course).[*]
Finally he extends his results to doubly infinite systems

$$(8) \qquad \sum_{k=-\infty}^{+\infty} a_{jk} x_k = b_j \qquad (-\infty < j < +\infty)$$

with the same restriction $a_{jj} = 1$ for all j; in particular
he shows that Hill's method is justified for the system (7)
([177], vol. V, p. 95-107).

Ten years later, H. von Koch [220] refined and generalized
Poincaré's results. Instead of making assumptions on the
diagonal terms, he writes the coefficients $\delta_{jk} + c_{jk}$ instead
of a_{jk} (with the Kronecker delta), and uses the expression
of a determinant $\Delta_n = \det(\delta_{jk} + c_{jk})_{1 \leq j, k \leq n}$ as a sum of principal minors

---

[*]One must beware of the fact that the Fourier method (when
no condition is imposed on the a_{jk}) may very well give convergent "infinite determinants", but the values given by the
Cramer formulas may be such that the left hand sides of (4)
are <u>divergent</u> series. An example is given by taking $a_{jk} = 0$
if $j > k$, $a_{jk} = 1$ if $j \leq k$, $b_j = (-1)^j$; one finds as a "solution" $x_k = 2(-1)^k$.

$$\Delta_n = 1 + \sum_{s=1}^{n} c_{ss} + \frac{1}{2!} \sum_{s_1, s_2} \begin{vmatrix} c_{s_1 s_1} & c_{s_1 s_2} \\ c_{s_2 s_1} & c_{s_2 s_2} \end{vmatrix} +$$

(9)

$$+ \frac{1}{3!} \sum_{s_1, s_2, s_3} \begin{vmatrix} c_{s_1 s_1} & c_{s_1 s_2} & c_{s_1 s_3} \\ c_{s_2 s_1} & c_{s_2 s_2} & c_{s_2 s_3} \\ c_{s_3 s_1} & c_{s_3 s_2} & c_{s_3 s_3} \end{vmatrix} + \ldots$$

(an expression which will be the starting point of Fredholm's theorems on integral equations 4 years later (Chap.V, §1)). He is thus able to replace Poincaré's criterion for convergence by a weaker one: it is enough that the sums $\sum_j |c_{jj}|$ and $\sum |c_{i_1 i_2} c_{i_2 i_3} \ldots c_{i_p i_1}|$ (extended to all sequences $(i_1, i_2, \ldots, i_p)$ of distinct indices) be finite. Another convergence criterion is that the sums $\sum_j |c_{jj}|$ and $\sum_{j,k} |c_{jk}|^2$ be finite.

§3 - Groping towards function spaces

It should not be believed that set-theoretic concepts in mathematics were unknown before Boole (1847) or Cantor; they can be traced at least as far back as Aristotle. The use of the word "class" (or, in German, "Gebiet", "Inbegriff", "Mannigfaltigkeit", "System") to designate a set of objects having a common property, becomes frequent among mathematicians since the beginning of the XIXth century. But it is only after Boole, in the second half of the century, that

using letters to denote more or less arbitrary sets, and computing with these letters, will become a widespread practice.

In particular "classes" of functions were very often considered in Analysis, even if their description lacks precision
most of the time. Even more widespread was the use, since the
XVIIIth century, of sequences of functions, or of functions
depending on one or several real parameters (for instance in
the Calculus of variations). It was of course dimly realized
that such families of functions were "much smaller" than the
"class" of all functions under consideration; the first attempt to give a clearer expression to that feeling is probably due to Riemann. In his famous inaugural lecture on the
foundations of geometry, after having tried to give an idea
of what he means by a "finite dimensional multiplicity (i.e.
manifold)" where the position of a point is determined by a finite set of numbers, he adds that there are "multiplicities"
(Mannigfaltigkeiten) for which such a determination is not
possible, but needs "an infinite sequence or a continuous multiplicity of numbers", and gives as an example "all the possible determinations of a function in a given domain" ([182],
p.276).

The extension of the concepts of limit and of continuity to
mathematical objects other than numbers or points, such as
curves, surfaces or functions, is also very old. However,
the applications of that idea dealt with sequences of such objects, or families depending on a finite number of real parameters; again, Riemann seems to have been the first to conceive that a whole "class" of functions might be given some
kind of "geometrical" structure (what we now would call a topo-

logy), for when he speaks of the functions for which the
Dirichlet integral (Chap. II, formula (21)) has a meaning, he
says that "this set of functions constitutes a connected domain, closed in itself" ([182], p.30), and although it is not
quite clear what he means by that, we may see in that statement a first glimpse of the notion of compactness, which will
emerge in the last part of the century (see below).

The rigorous study of limits of sequences of functions, which began around 1820, brought to light a phenomenon
which had no counterpart for sequences of numbers or of points
in $\mathbb{R}^n$: there are **several distinct ways** for a sequence (f_n)
of functions to tend to a limit f. The first problem occur-
red with the distinction between simple and uniform convergence, which was only quite cleared up around 1850. This was
followed in the last third of the XIXth century by a deeper
study of these notions, chiefly due to the Italian school
(Dini, Ascoli, Arzelà); the most important step taken by that
school was the introduction by Ascoli in 1883 of the notion of
equicontinuity. He discovered that the unpleasant phenomenon
of a sequence of continuous functions (in a bounded closed interval I), converging simply to a discontinuous function,
would disappear if one assumed on the sequence the following
additional property: for each $\varepsilon > 0$, there exists a $\delta > 0$
such that, if $|x'-x''| \leq \delta$, then $|f_n(x') - f_n(x'')| \leq \varepsilon$ for
all indices n (in other words, the continuity is "uniform",
not only with respect to x, but also with respect to n)
[8].

One of the fundamental properties of equicontinuous sequences is that, when in addition the f_n are uniformly bounded,

it is possible to find a <u>subsequence</u> (f_{n_k}) which converges
uniformly, a generalization of the "Bolzano-Weierstrass"
theorem for sequences of numbers, which was well-known after
1880. This "compactness" property (which holds for functions
defined in a closed bounded set of $\mathbb{R}^n$) was thrust in the li-
melight by Hilbert, who apparently rediscovered it independently in a special case (he does not quote the Italians) and used
it as an essential tool in his famous 1900 paper where he invented the "direct method" in the Calculus of variations
([111], vol. III, p.10-14) and thus was able to justify
Riemann's use of the "Dirichlet principle" (chap.II, §3)
(<u>loc.cit.</u>, p.15-37).

It is also from the Calculus of variations that another notion of "neighborhood" for a function emerged during the last
years of the XIXth century. Already at the end of the XVIIIth
century mathematicians investigated the problem of deciding if
a solution y of the Euler equation for an integral
$\int_a^b F(x,y,y')dx$ actually gave a "relative extremum" for that
integral. Legendre tried to give a solution to that problem
by replacing y in the integral by $y + \varepsilon u$,   where $u = \delta y$
is an arbitrary "variation" of class C^1; he thus obtains a
function $\Phi(\varepsilon)$ of the real parameter ε and if $\Phi''(0) > 0$
(resp. $\Phi''(0) < 0$) that function reaches a relative minimum
(resp. maximum) for $\varepsilon = 0$. This yields the condition $\dfrac{\partial^2 F}{\partial y'^2} > 0$
(resp. < 0); but it was soon realized that this condition was
not sufficient to guarantee that the integral would actually
be smaller (resp. larger) than <u>all</u> numbers obtained by replacing y by $y + \delta y$ for a "small" variation δy. Clearly this

hinges on the question of what exactly is meant by the word "small". Ever since Lagrange, it had been taken for granted that the derivative $(\delta y)' = \delta y'$ is "small" whenever δy itself is "small"; but Weierstrass and his school realized that this was an additional assumption, and this led them to distinguish between "strong extremum" and "weak extremum": the second corresponds to a notion of "neighborhood" of a C^1 function y, where z is "close" to y when the maximum of $|z-y|$ is small, whereas for the first z is only considered as "close" to y if <u>both</u> the maximum of $|z-y|$ and the maximum of $|z'-y'|$ are small.

Finally, we have noticed earlier that Gram and Poincaré were naturally confronted with the notion of "convergence in the mean square" in their study of "Fourier expansions" (chap.III, §2). We may therefore say that in the last years of the XIXth century, the idea of "function spaces" with various "topologies" was so to speak "in the air", and ready to blossom forth as soon as it could be expressed in sufficiently general and simple terms.$^{(*)}$

The concept of <u>mapping</u> of a set of functions into $\mathbb{R}$, or

---

$^{(*)}$It is, however, typical of the unpredictability of mathematical developments that nobody seems to have been able to foresee, even conjecturally, the direction which was taken by Functional Analysis in the fateful years 1900-1910. This is clear in the communication made by Hadamard in the first International Congress of mathematicians in 1897 ([94], vol.I, p.311-312); he was keenly interested in these "set-theoretical" ideas, and had great expectations of what was to come; but he could think of no serious applications beyond the rehabilitation of the "Dirichlet principle" and some vague ideas on what we now call "precompactness".

into another set of functions, is also much older than the general definition of a mapping of an arbitrary set into an arbitrary set, which does not seem to have been formulated before Dedekind's famous "Was sind und was sollen die Zahlen",
written in 1872 (although only published in 1888) ([48], vol.
III, p.335-391). Ever since the beginning of the Calculus of
variations, mathematicians were familiar with the idea of at-
taching for instance to each C^1 function y in an interval
$[a,b]$ a number $\int_a^b F(x,y,y')dx$ depending on y; such mappings would receive the name of "functional" at the end of
the XIXth century. Similarly, as soon as the concept of function emerged at the end of the XVIIth century together with
its use in Calculus, the concept of operator, yielding a new
function when applied to a given function, was in evidence
with the examples of the derivatives $f \mapsto D^\nu f$ or the translation operator $f \mapsto \gamma(a)f$ (function $x \mapsto f(x-a)$); and from
Leibniz to Pincherle (end of the XIXth century) many analysts
were led to ponder on the algebraic properties of these operators, and their similarity with results of ordinary algebra
(which was originally conceived as applying to numbers only).
For instance, the similarity of Leibniz's formula for the
iterated differential $d^n(uv)$ of a product, with the binomial
theorem, probably gave him the idea of attempting to introduce
differentials d^α with negative or irrational exponents, a
problem to which many mathematicians (such as Liouville,
Riemann, Pincherle) later returned, and which has only finally been put to rest with the modern theory of distributions.
Other examples are the expression of Taylor's formula given

by Lagrange as a relation $\gamma(-a) = e^{aD}$ between operators, or
the factoring of a differential polynomial $D^n + a_1 D^{n-1} + \ldots + a_n$
on the model of the factoring of an ordinary polynomial
$z^n + a_1 z^{n-1} + \ldots + a_n$.

Such ideas, abundantly developed in the period 1790-1830,
had much to do with the new conception of Algebra as dealing
with symbols rather than with numbers, and later with the
axiomatic and formalist conception of the whole of mathematics
(see [54], chap. XIII, §III); but they had no perceptible influence on Analysis, probably because they did not pay much
attention to questions of continuity. It is only in the last
years of the XIXth century that such questions appear, in a
very episodic way, in papers by Pincherle, Bourlet and Volterra.

The first two of these authors only consider one "space" E,
the set of all holomorphic functions in a domain Δ of the
complex plane, and they are exclusively concerned with <u>linear</u>
operators in that space. In 1886, Pincherle studies operators
which, to a holomorphic function[(*)] φ, associate the function
$x \mapsto \int_\Gamma A(x,y)\varphi(y)dy$, where Γ is a curve in Δ and A is
holomorphic, and he writes that function $\mathfrak{a}\varphi$, but he limits
himself to special cases, of the type of the Laplace transform
([173], vol.I, p.92-141). He several times returned later to

[(*)] After Grassmann (1862), Pincherle seems to have been one
of the first mathematicians to write a function with a single
letter φ, when all his contemporaries wrote $\varphi(x)$. In his
later papers, he repeatedly insists on the fact that a function should be considered as a "point" in some set.

such questions, but failed to obtain any substantial results[*].
In 1897, Bourlet [29], limiting himself to the case in which
Δ is a disk $|z| < r$, explicitly determines the linear operators in E which are "continuous" (by which he means continuity for what we now call the topology of compact convergence), showing that they are integral operators of the form considered by Pincherle.

We must finally mention the first attempts at "Functional
Analysis" of the young Volterra in 1887 ([219], vol.I, p.
294-314), to which, under the influence of Hadamard, has been
attributed an exaggerated historical importance. Volterra
had in mind a generalization of analytic functions, which may
be considered as a prefiguration of Hodge's theory[**]; for
this he needs what he calls "functions of lines". Although,
from our point of view, his definitions are not very precise[***],
he apparently considers the set E of C^1 mappings of an

---

[*] He should however be credited with what is probably the
first conception of a closed hyperplane in E as the kernel of
a continuous linear form, and of closed subspaces of finite
codimension as intersections of hyperplanes ([173], vol. I,
p.395). In 1897-98, he also has the idea of generalizing
Lagrange's "adjoint" of an operator (chap.I, §1, formula (5))
by considering two vector subspaces S, S' of E, and a nonde-
generate bilinear form (φ, ψ) on S×S'; to a linear mapping A
of S into S', he then associates the "adjoint" A', a linear
mapping of S' into S such that $(A \cdot \varphi, \psi) = (\varphi, A' \cdot \psi)$, and he
observes the relation between the kernel of A and the image
of A' ([173], vol.II, p.77-84).

[**]
    See A. Weil, Oeuvres Scientifiques, vol.II, Commentaires
sur [1952 e], p.532 of the corrected edition (or vol.III, p.450
of the first printing), Springer, Berlin-Heidelberg-New York,
1979.

[***]
    This can be said of practically all mathematicians
before 1906.

interval $I \subset \mathbb{R}$ into $\mathbb{R}^3$ (the "lines"), and the mappings
$y: E \rightarrow \mathbb{R}$, continuous for the topology of uniform convergence.
For these "functions of lines" he immediately wants to gene-
ralize the classical notion of derivative; in a manner reminiscent of the Calculus of variations, he considers a "varia-
tion" $\delta y = y(\varphi+\theta) - y(\varphi)$, where the increment θ is supposed to vanish outside of an interval $[a,b]$, and then the
quotient $\delta y/\sigma$, where $\sigma = \int_a^b |\theta(t)| dt$; this should tend to
a limit when $b-a$ and the maximum of $|\theta|$ tend to 0. With
our experience of 50 years of Functional Analysis, we cannot
help feeling that, without even the barest notions of general
topology, these ad hoc definitions were decidedly premature.
Nevertheless, they caught the fancy of Hadamard, who tried to
apply similar ideas to Green's functions and encouraged his
students to work in that direction (see [94, vol.I, p.401-404
and 435-453] and [146]). But these ideas have not, up to now,
produced anything comparable to the applications of spectral
theory and distribution theory, which we will describe in
chap. VII and IX; it might be worthwhile to reexamine them in
the light of recent progress in the theory of infinite dimensional manifolds, which could be their natural setting.

§4 - The passage "from finiteness to infinity"

The urge to deal with "infinity" has been present from the
very beginnings of Greek mathematics, in spite of all philo-
sophical preconceptions and objections, and has taken various
forms. The simplest and most "natural" passage "from finite-

ness to infinity" is the "indefinite repetition" of the arith-
metical operation of addition, on smaller and smaller summands,
giving birth to the concept of convergent series, of which one
can already find examples in Archimedes. Replace addition by
multiplication, and you have the infinite product, born with
Calculus in the XVIIth century; and still more sophisticated
algebraic manipulations would lead to continued fractions and
to the infinite determinants which we have discussed in §2.

Another line of thought goes back at least to Eudoxus's
"method of exhaustion", and was to lead in the first place to
the concept of integral. But in the hands of the mathematicians of the XVIIth and XVIIIth century, this idea of decomposing an object into "infinitesimal" parts in which the phenomenon they studied became much easier to describe "in a
first approximation", was developed into a more and more sophisticated method to discover the differential or partial
differential equations which governed the phenomenon "in the
large". It is in that way that the equation of vibrating
strings (chap.I, §2, formula (7)) was established, either by
considering, as D. Bernoulli, a massive string as a limit
(for n tending to infinity) of a system of n massive points
distributed on a massless string, or by analyzing, as d'Alembert, the forces which are exerted on an "infinitesimal" portion of the string by its neighbors.

It is this second method that Fourier applied to obtain the
heat equation; he takes for granted that in a system of small
"molecules", a given molecule M receives in an "infinitesimal" time dt a quantity of heat from another molecule M′
equal to the difference of temperatures of M and M′, mul-

tiplied by dt and by a coefficient depending only on the
distance MM′; the molecule M, if situated at the surface
separating the system of molecules from the external world,
also radiates a quantity of heat equal to the difference of
its temperature and of the external temperature, multiplied
by dt and another coefficient depending on M. He then derives the equation of the "cooling off" process (chap.III,
§2, equation (21)) by decomposing the solid body V in "infinitesimal" cubes and evaluating the amount of heat received
by one of them from its 6 neighbors in time dt, which he
takes as proportional (with a constant coefficient) to the
variation du of the temperature of that cube; the boundary
condition (chap.III, §2, equation (22)) is similarly obtained
by evaluating the amount of heat lost (by radiation) by an infinitesimal cube at the surface of V.

   At the end of his 1890 paper on the cooling off problem
(chap.III, §2), Poincaré suggests another method reminiscent
of D. Bernoulli's procedure. He first considers a large number N of molecules M_i; following Fourier's physical consi-
derations, and denoting by $v_i(t)$ the temperature of M_i at
time t, these functions satisfy the system of linear differential equations

$$(10) \qquad \frac{dv_i}{dt} + \sum_{k \neq i} C_{ik}(v_i - v_k) + C_i v_i = 0 \qquad (1 \le i \le N),$$

$C_{ik}(v_i - v_k)$ being the quantity of heat received from M_k and
$C_i v_i$ the quantity of heat radiated by M_i outside the system.
But instead of letting the number of molecules increase to infinity, Poincaré _first_ integrates the system (10) by the classical Euler-Lagrange method: he writes $v_i(t) = u_i e^{-\alpha t}$, and,

using the fact that the matrix (C_{ik}) is <u>symmetric</u>, he re-
cognizes in the equation he obtains for α the equation
giving the <u>eigenvalues</u> of the symmetric matrix corresponding
to the non degenerate positive <u>quadratic form</u>

(11) $\Phi(u_1, u_2, \ldots, u_N) = \sum_{i \neq k} C_{ik}(u_i - u_k)^2 + \sum_i C_i u_i^2$.

Let $\xi_1 \leq \xi_2 \leq \ldots \leq \xi_N$ be these eigenvalues; the classical
theory of quadratic forms shows that one may write

(12) $\Phi = \xi_1 \varphi_1^2 + \ldots + \xi_N \varphi_N^2$

where the φ_i are linear forms in the variables $u_1, \ldots, u_N$
such that $\varphi_1^2 + \ldots + \varphi_N^2 = u_1^2 + \ldots + u_N^2$; if for two such forms
$f = \alpha_1 u_1 + \ldots + \alpha_N u_N$, $g = \beta_1 u_1 + \ldots + \beta_N u_N$ one writes $(f|g) =$
$= \sum_k \alpha_k \beta_k$, the N forms φ_i are mutually orthogonal for that
scalar product. It is then clear that ξ_1 is the smallest
value of the function of $u_1, \ldots, u_N$

(13) $\dfrac{\Phi(u_1, \ldots, u_N)}{u_1^2 + u_2^2 + \ldots + u_N^2} = \dfrac{\Phi}{\varphi_1^2 + \ldots + \varphi_N^2}$

where the u_i are arbitrary; similarly ξ_2 is the minimum
of (13) for $\varphi_1 = 0$, ξ_3 the minimum for $\varphi_1 = \varphi_2 = 0$ as
relations between the u_i; and so on. This is of course the
analogous procedure in N dimensions to the classical determination of the "axes" of an ellipsoid in 3-dimensional space.
Poincaré's idea is that the expression (13) corresponds exactly to the quotient

$$\dfrac{\iiint_V \left(\left(\frac{\partial v}{\partial x}\right)^2 + \left(\frac{\partial v}{\partial y}\right)^2 + \left(\frac{\partial v}{\partial z}\right)^2 \right) d\omega + h \iint_\Sigma v^2 d\sigma}{\iiint_V v^2 d\omega}$$

in his (or rather Weber's) procedure for the definition of the
eigenvalues in the cooling off problem; these eigenvalues (the
poles λ_m of his function $[f,\xi]$ (chap.III, §2)) correspond
to the ξ_j in (12) and the eigenfunctions $U_m(M) = P(M,\lambda_m)$
to the φ_j, the orthogonality of the φ_j corresponding to
the relations

$$\iiint_V U_p U_q \, d\varpi = 0 \quad \text{for} \quad p \neq q$$

between the U_m. Finally, he realizes that the same ideas
apply as well to other problems and gives as an example the
theory of elasticity, and he suggests that a rigorous proof
of the existence of the λ_m and the U_m, which he had not
been able to give, might be obtained by simply letting N
tend to $+\infty$ in the formula (12). He never came back to the
question; but we cannot fail to see that this is exactly the
program which Hilbert in 1904 followed to its successful conclusion for integral equations with symmetric kernel (chap.V,
§2).

A similar "passage from finiteness to infinity" emerged in
the first general theory of integral equations, beginning with
the papers of Le Roux in 1894 and Volterra in 1896. In addition to the particular integral equations which had been met
by Liouville in the Sturm-Liouville problem (chap.I, §3, equation (34) and chap.II, §1, equation (6)) and by Beer and
Neumann in the Dirichlet problem (chap.II, §4, equation (23))
(not to speak of what we have called "crypto-integral" equations, where the equation is not written down explicitly but
the method exactly amounts to solving it), other particular
equations involving integrals had come up in connection with

problems not directly related to differential or partial differential equations. The first one (chronologically) was the
"inversion" problem for the "transform" introduced by Fourier
in 1822 (and to which we shall return in chap. VII, §6); it
associates to a function f in $[0, +\infty[$ the function

$$(14) \qquad\qquad \varphi(t) = \int_0^{+\infty} f(x)\cos tx\ dx$$

and the problem consisted in finding f when the transform
φ is a given function. It was solved by Fourier's inversion
formula ($[67]$, vol.I, p.392)

$$(15) \qquad\qquad f(x) = \frac{2}{\pi} \int_0^{\infty} \varphi(x)\cos tx\ dt$$

where, as usual with Fourier, both formulas are obtained by
a purely formal calculation. A little later, one of the
first published papers of Abel ($[1]$, vol.I, p.11-27 and
97-101) was devoted to a problem of mechanics, which amounted
to finding a function φ such that

$$(16) \qquad\qquad \int_0^x \frac{\varphi(y)dy}{\sqrt{x-y}} = \psi(x)$$

is a given function; he obtains the solution by the formula

$$(17) \qquad\qquad \varphi(x) = \frac{1}{\pi} \int_0^x \frac{\psi'(y)dy}{\sqrt{x-y}}$$

and extends his result to the case in which $\sqrt{x-y}$ is replaced by $(x-y)^{\alpha}$ for $0 < \alpha < 1$. In a letter to Holmboe, he
even hinted at more general results, but nothing was found on
the subject in his papers. After Abel, a few papers, giving
partial generalizations of his results, were published until

$1890^{(*)}$; but it was only in 1894 that Le Roux attacked the
general problem of "inversion of a definite integral" (as it
was called), i.e. finding a C^1 function φ in an interval
$[a,b]$ satisfying an equation

(18)
$$\int_a^y \varphi(x)H(x,y)dx = f(y)$$

where f and H are C^1 (in $[a,b]$ and $[a,b] \times [a,b]$
respectively) and $f(a) = 0$ $^{(**)}$. In contrast with his predecessors, Le Roux is not trying to find a "closed formula"
similar to (15) and (17) for the unknown function. He assumes that $h(y) = H(y,y)$ does not vanish in $[a,b]$, takes the
derivative of both sides of (18), obtaining

(19)
$$h(y)\varphi(y) + \int_a^y \frac{\partial H}{\partial y}(x,y)\varphi(x)dx = f'(y)$$

and then applies the method of successive approximations which
Picard had popularized a few years earlier:

$$u_o(y) = \frac{f'(y)}{h(y)}, \quad u_n(y) = \frac{f'(y)}{h(y)} - \frac{1}{h(y)}\int_a^y \frac{\partial H}{\partial y}(x,y)u_{n-1}(x)dx$$
$$\text{for } n \geq 1,$$

proving easily the convergence of the sequence (u_n) to a
solution of (18) ($[143]$, p.244-246).

In 1896, Volterra (who apparently was unaware of Le Roux's
paper) tackles exactly the same problem by the same method,

$^{(*)}$See the long historical introduction given by Volterra in
his 1897 paper on integral equations ($[219]$, vol.II,p.279-287).

$^{(**)}$As these conditions are not satisfied for Abel's equation,
Le Roux's results (which for him are auxiliary properties
which he needs in a study of partial differential equations)
do not directly generalize those of Abel.

in a series of 4 notes ([219], vol.II, p.216-262). He goes
a little beyond Le Roux, by giving an explicit expression of
the solution

$$(20) \qquad \varphi(y) = \frac{f'(y)}{h(y)} - \frac{1}{h(y)} \int_a^y (\sum_{i=0}^{\infty} S_i(x,y)) f'(x) dx$$

where the S_i are defined by induction:

$$(21) \quad S_o(x,y) = \frac{1}{h(x)} \frac{\partial H}{\partial y}(x,y), \quad S_i(x,y) = \int_y^x S_o(\xi,y) S_{i-1}(x,\xi) d\xi$$

$$\text{for} \quad i \geq 1.$$

In the later notes, he discussed the cases in which $h(y)$ may
vanish at a finite number of points, and the case in which
$H(x,y) = G(x,y)/(x-y)^{\alpha}$ with $0 < \alpha < 1$ and G is continuous
(the generalization of Abel's equation). But the most influential part of his notes was the following remark he made
immediately after obtaining formula (20): "If one considers
the system

$$(22) \qquad \begin{cases} b_1 = a_{11}x_1 \\ b_2 = a_{12}x_1 + a_{22}x_2 \\ \cdot \quad \cdot \quad \cdot \quad \cdot \quad \cdot \quad \cdot \\ b_n = a_{1n}x_1 + \ldots + a_{nn}x_n \end{cases}$$

the concept of integral easily leads to look at the question
of functional Analysis represented by equation (18) as a limiting case of the solution of a system similar to (22), in
which the a_{ij} and a_{ii} are the analogous of $H(x,y)$ and
$H(y,y)$." Although he limited himself to that (somewhat vague)
statement, it seems obvious that what he had in mind was replacing in (18) the variable y by its values

$$y_k = a + \frac{k}{n}(b-a) \qquad \text{for} \qquad 1 \leq k \leq n$$

and replacing the integral by the corresponding "Riemann sum" for the subdivision of $[a,b]$ by the points y_k, obtaining the system of type (22)

$$f(y_j) = \frac{b-a}{n} \sum_{k=1}^{n} \varphi(y_k) H(y_k, y_j) \quad \text{for} \quad 1 \leq j \leq n.$$

Finally, although he does not mention the product of matrices, Volterra develops in these notes the formalism which to two "kernels" $H_1(x,y)$, $H_2(x,y)$ associates the kernel

$$H(x,y) = \int_x^y H_1(x,\xi) H_2(\xi,y) d\xi$$

(which much later he will write $H = H_1 * H_2$). If, for simplicity, we adopt this notation, he shows that, for an <u>arbitrary</u> continuous function $S_o(x,y)$ if we define for $i \geq 1$ the "kernel" S_i by $S_i = S_{i-1} * S_o$, one has also $S_i = S_{i-j} * S_{j-1}$ for $1 \leq j \leq i$, and a majoration

$$|S_i(x,y)| \leq \frac{M^{i+1}}{i!} |x-y|^i .$$

This implies uniform convergence for the series

$$(23) \qquad\qquad F_o = \sum_{i=0}^{\infty} S_i$$

and the relation

$$(24) \qquad\qquad F_o - S_o = S_o * F_o .$$

He observes that one may "invert" that relation: if the F_i are defined for $i \geq 1$ by $F_i = F_o * F_{i-1}$, one has

$$(25) \qquad\qquad S_o = \sum_{i=0}^{\infty} (-1)^i F_i .$$

And finally, at the end of his notes, he arrives at the general concept of what Hilbert will call an "integral equation

of the second kind"

$$(26) \qquad \varphi(y) - \int_a^y S_o(x,y)\varphi(x)dx = f(y)$$

for which the solution is given by

$$(27) \qquad \varphi(y) = f(y) + \int_a^y F_o(x,y)f(x)dx$$

as it follows immediately from (24), the "kernel" and the

"resolvent kernel" playing completely symmetric parts in these

formulas.

CHAPTER V

THE CRUCIAL YEARS AND THE DEFINITION OF HILBERT SPACE

Between 1900 and 1910, there was a sudden crystallization of all the ideas and methods which had been slowly accumulating during the XIXth century and which we have described in the previous chapters. This was essentially due to the publication of <u>four fundamental papers</u>:

Fredholm's 1900 paper on integral equations;

Lebesgue's thesis of 1902 on integration;

Hilbert's paper of 1906 on spectral theory;

Fréchet's thesis of 1906 on metric spaces.

§1 - Fredholm's discovery

The name "integral equation" (<u>Integralgleichung</u>) was used for the first time by P. du Bois-Reymond in 1888, in a paper on the Dirichlet problem [61]; he has in mind equations of the Beer-Neumann type (chap.II, §4) and considers that a general theory of such equations presents "insuperable difficulties"; he is convinced that much progress would come out of such a theory but acknowledges that "almost nothing is known on this question". The later work of Poincaré, which we have discussed above (chap.III, §2), and of his immediate followers, did nothing to dispel that impression; their results seemed linked

97

to delicate estimates from potential theory. It therefore
came as a complete surprise when, in a short Note published
in 1900, Fredholm showed that the general theory of all integral equations (or "crypto-integral" equations) considered before him was in fact extremely simple (much simpler than anything known at the time in the theory of partial differential
equations).

Ivar Fredholm (1866-1927) was a student of Mittag-Leffler in
Stockholm in 1888-1890; he only published a few papers during
his lifetime, mostly concerned with partial differential
equations (we shall return to his thesis of 1898 in chapter
IX, §5). After a visit to Paris, where he had been in contact with all the French analysts and had become familiar with
the recent papers of Poincaré, he communicated in August 1899
his first results on integral equations to his former teacher;
they were published in 1900 [74, p.61-68] and completed 2 years
later in a paper published in <u>Acta Mathematica</u> ([74, p.81-106]
and [75]).

Fredholm's 1900 note is entitled "On a new method for the
solution of Dirichlet's problem", but it is characteristic
that from the start, he brushes aside all the particular features of the Beer-Neumann equation, and (as Le Roux and Volterra had done with Abel's equation (chap.IV, §4)) begins with
a <u>general</u> "integral equation of the second kind" (that name
will only be given by Hilbert)

$$(1) \qquad \varphi(s) = f(s) + \lambda \int_a^b K(s,t)f(t)dt$$

where K is supposed to be bounded and piecewise continuous

in $[a,b] \times [a,b]$, and φ continuous in $[a,b]$, λ being a complex parameter. He briefly mentions the analogy with systems of linear equations and starts right away with the formulas describing his "determinants" (see below). But in a lecture given in 1909 [74, p.123-131], he acknowledges: 1º the inspiration derived from Volterra's idea of a "passage to the limit" from a system of linear equations to an integral equation; 2º the help he found in von Koch's work on infinite determinants (chap.IV, §2). From these indications, sparse as they are, it seems one can reconstruct his procedure, with great probability, as consisting in putting together three simple ideas:

I) Replacing the integral in (1) by Riemann sums, one obtains, with the notations of chap. IV, §4, the system of n linear equations for the $f(y_j)$

$$(2) \quad f(y_j) + \frac{\lambda(b-a)}{n} \sum_{k=1}^{n} K(y_k, y_j) f(y_k) = \varphi(y_j) \quad (1 \le j \le n).$$

II) Writing the determinant of that system according to von Koch's formula (chap.IV, §2, formula (9))

$$1 + \frac{\lambda(b-a)}{n} \sum_{k=1}^{n} K(y_k, y_k) + \frac{\lambda^2(b-a)^2}{2!\,n^2} \sum_{k_1,k_2} \begin{vmatrix} K(y_{k_1}, y_{k_1}) & K(y_{k_1}, y_{k_2}) \\ K(y_{k_2}, y_{k_1}) & K(y_{k_2}, y_{k_2}) \end{vmatrix} + \ldots$$

and then letting n tend to $+\infty$, which gives the formula for what Fredholm calls the "determinant" of the integral equation (1)

$$\Delta(\lambda) = 1+\lambda \int_a^b K(s,s)ds + \frac{\lambda^2}{2!} \int_a^b \int_a^b K\begin{pmatrix} s_1 & s_2 \\ s_1 & s_2 \end{pmatrix} ds_1\, ds_2 + \cdots$$

(3)
$$+ \frac{\lambda^m}{m!} \int_a^b \cdots \int_a^b K\begin{pmatrix} s_1 & s_2 & \cdots & s_m \\ s_1 & s_2 & \cdots & s_m \end{pmatrix} ds_1\, ds_2\, \cdots\, ds_m + \cdots$$

where he has written

(4)
$$K\begin{pmatrix} x_1 & x_2 & \cdots & x_m \\ y_1 & y_2 & \cdots & y_m \end{pmatrix} = \begin{vmatrix} K(x_1,y_1) & K(x_1,y_2) & \cdots & K(x_1,y_m) \\ K(x_2,y_1) & K(x_2,y_2) & \cdots & K(x_2,y_m) \\ \cdots & \cdots & \cdots & \cdots \\ K(x_m,y_1) & K(x_m,y_2) & \cdots & K(x_m,y_m) \end{vmatrix}.$$

III) Proving the uniform convergence of the series (3) in any compact set of the complex plane, for which it is enough to majorize the determinants (4) in a suitable way; in his 1899 letter, Fredholm had given the majoration $n^{n/2}M^n$, where M is the upper bound of $|K|$; he had apparently arrived independently to this result, but was made aware that it was a special case of an inequality published by Hadamard in 1893 ([94], vol.I, p.239-245) for an arbitrary square matrix $A = (a_{ij})$ of order n:

(5)
$$|\det(A)|^2 \le \prod_{i=1}^n \left(\sum_{j=1}^n |a_{ij}|^2 \right).$$

The next "natural" steps are of course to apply Cramer's formulas to the system (2) and let again n tend to infinity in the numerators; the result is described by Fredholm in the following elegant way: a development of the determinant (4) according to the first row yields the formula

$$K\begin{pmatrix} s & x_1 & \cdots & x_m \\ t & x_1 & \cdots & x_m \end{pmatrix} = K(s,t) \, K\begin{pmatrix} x_1 & \cdots & x_m \\ x_1 & \cdots & x_m \end{pmatrix} -$$

$$(6) \quad -K(s,x_1)K\begin{pmatrix} x_1 & x_2 & \cdots & x_m \\ t & x_2 & \cdots & x_m \end{pmatrix} + K(s,x_2)K\begin{pmatrix} x_1 & x_2 & x_3 & \cdots & x_m \\ t & x_1 & x_3 & \cdots & x_m \end{pmatrix} -$$

$$- \cdots + (-1)^m K(s,x_m)K\begin{pmatrix} x_1 & x_2 & \cdots & x_m \\ t & x_1 & \cdots & x_{m-1} \end{pmatrix}.$$

On the other hand, Fredholm defines the "minor"

$$(7) \quad \begin{aligned} \Delta(s,t;\lambda) &= K(s,t) + \lambda \int_a^b K\begin{pmatrix} s & x_1 \\ t & x_1 \end{pmatrix} dx_1 + \cdots + \\ &+ \frac{\lambda^m}{m!} \int_a^b \cdots \int_a^b K\begin{pmatrix} s & x_1 & \cdots & x_m \\ t & x_1 & \cdots & x_m \end{pmatrix} dx_1 dx_2 \cdots dx_n + \cdots \end{aligned}$$

and replaces each integrand by its expression (6), which gives him the simple relation

$$(8) \qquad \Delta(s,t;\lambda) = K(s,t)\Delta(\lambda) - \lambda \int_a^b K(s,\xi)\Delta(\xi,t;\lambda)d\xi.$$

He then introduces the function

$$(9) \qquad \Phi(s) = \varphi(s)\Delta(\lambda) - \lambda \int_a^b \Delta(s,\xi;\lambda)\varphi(\xi)d\xi$$

and derives from (8) the equation

$$(10) \qquad \Phi(s) + \lambda \int_a^b K(s,t)\Phi(t)dt = \varphi(s)\Delta(\lambda).$$

The conclusion is then immediate: if $\Delta(\lambda) \neq 0$, the function $f(s) = \Phi(s)/\Delta(\lambda)$ is a solution of (1). Furthermore, he shows that one has

$$(11) \qquad \frac{d\Delta(\lambda)}{d\lambda} = \int_a^b \Delta(s,s;\lambda)ds$$

and from this he deduces that if λ_o is a zero of order ν

of the entire function $\Delta(\lambda)$, $\Phi(s)$, for a suitable choice of φ, cannot be divisible by a power of $\lambda - \lambda_o$ greater than $(\lambda - \lambda_o)^{\nu - 1}$; if $\Phi(s) = (\lambda - \lambda_o)^k \Phi_1(s)$, one then deduces, from (10), that

$$(12) \qquad \Phi_1(s) + \lambda_o \int_a^b K(s,t)\Phi_1(t)dt = 0;$$

in other words, if there is no nontrivial solution of the homogeneous equation (12), necessarily $\Delta(\lambda_o) \neq 0$, hence the solution of (1) for $\lambda = \lambda_o$ exists and is unique. However, at that time, he does not yet prove that the existence of a non trivial solution of (12) implies that $\Delta(\lambda_o) = 0$. But the end of the Note is startling: he considers the Beer-Neumann equation for a bounded plane domain with a C^3 boundary; the kernel of that integral equation is then bounded and continuous, and for $\lambda_o = 1$ it is very easy to deduce from the properties of double layer potentials that the homogeneous equation (12) has no nontrivial solution. Therefore the existence and uniqueness of the solution of Dirichlet's problem is proved, doing away, with a single stroke of the pen, so to speak, with all the complications of the Neumann-Poincaré solution!

In his 1903 paper, Fredholm completed his results on some important points. He first defines more general "minors"

$$\Delta \begin{pmatrix} s_1 & s_2 & \cdots & s_m \\ t_1 & t_2 & \cdots & t_m \end{pmatrix} ; \lambda) = K \begin{pmatrix} s_1 & s_2 & \cdots & s_m \\ t_1 & t_2 & \cdots & t_m \end{pmatrix} +$$

$$(13)$$

$$+ \sum_{n=1}^{\infty} \frac{\lambda^n}{n!} \int_a^b \cdots \int_a^b K \begin{pmatrix} s_1 & \cdots & s_m & x_1 & \cdots & x_n \\ t_1 & \cdots & t_m & x_1 & \cdots & x_n \end{pmatrix} dx_1 \cdots dx_n .$$

Developing this time the determinants both according to the

first row and the first column, he obtains the identities

$$\Delta \begin{pmatrix} s_1 & \cdots & s_m \\ t_1 & \cdots & t_m \end{pmatrix};\lambda) + \lambda \int_a^b K(s_1,\xi)\Delta \begin{pmatrix} \xi & s_2 & \cdots & s_m \\ t_1 & t_2 & \cdots & t_m \end{pmatrix};\lambda)d\xi =$$

(14)

$$= K(s_1,t_1)\Delta \begin{pmatrix} s_2 & \cdots & s_m \\ t_2 & \cdots & t_m \end{pmatrix};\lambda) - K(s_1,t_2)\Delta \begin{pmatrix} s_2 & s_3 & \cdots & s_m \\ t_1 & t_3 & \cdots & t_m \end{pmatrix};\lambda)+\cdots$$

and

$$\Delta \begin{pmatrix} s_1 & \cdots & s_m \\ t_1 & \cdots & t_m \end{pmatrix};\lambda) + \lambda \int_a^b K(\xi,t_1)\Delta \begin{pmatrix} s_1 & s_2 & \cdots & s_m \\ \xi & t_2 & \cdots & t_m \end{pmatrix};\lambda)d\xi =$$

(15)

$$= K(s_1,t_1)\Delta \begin{pmatrix} s_2 & \cdots & s_m \\ t_2 & \cdots & t_m \end{pmatrix};\lambda) - K(s_2,t_1)\Delta \begin{pmatrix} s_1 & s_3 & \cdots & s_m \\ t_2 & t_3 & \cdots & t_m \end{pmatrix};\lambda)+\cdots$$

which in particular, for $m = 1$, reduce to (8) and to

$$(16) \qquad \Delta(s,t;\lambda) = K(s,t)\Delta(\lambda) - \lambda \int_a^b K(\xi,t)\Delta(s,\xi;\lambda)d\xi .$$

The use he makes of these formulas is a little more sophisticated than in his first Note. He introduces the operator
corresponding to the kernel K, $f \mapsto S_K f$ such that
$S_K f(s) = f(s) + \int_a^b K(s,t)f(t)dt$, and, for two kernels K, K',
writes the composite $S_K S_{K'}$ as $S_{K''}$ with

$$(17) \qquad K''(x,t) = K(x,t) + K'(s,t) + \int_a^b K(s,\xi)K'(\xi,t)d\xi .$$

Suppose now that $\Delta(\lambda) \neq 0$, and write

$$(18) \qquad R(s,t;\lambda) = -\Delta(s,t;\lambda)/\Delta(\lambda)$$

(the <u>resolvent kernel</u> in the later terminology of Hilbert).
It then follows from (17), (8) and (16) that we have

(19) $$S_{\lambda K} S_R = S_R S_{\lambda K} = Id$$

and Fredholm has thus shown that the necessary and sufficient condition for the existence and uniqueness of a solution of (1) is $\Delta(\lambda) \neq 0$, the kernel λK and the resolvent kernel R playing completely symmetric parts as in the formulas of Volterra (chap. IV, §4, formulas (26) and (27)).

Next he examines what happens when $\Delta(\lambda) = 0$. First he generalizes (11) to

(20) $$\frac{d^m \Delta(\lambda)}{d\lambda^m} = \int_a^b \cdots \int_a^b \Delta\left(\begin{matrix} s_1 & \cdots & s_m \\ s_1 & \cdots & s_m \end{matrix}; \lambda\right) ds_1 \cdots ds_m$$

and from this he deduces that if $\Delta(\lambda) = 0$, there is always an integer m such that $\Delta\left(\begin{matrix} s_1 & \cdots & s_m \\ t_1 & \cdots & t_m \end{matrix}; \lambda\right)$ is not identically 0. If m is the smallest integer having that property (which is exactly the order of λ as a zero of Δ) he exhibits, using (14), m solutions of the homogeneous equation

(21) $$\Phi_1(s) = \frac{\Delta\left(\begin{matrix} s & s_2 & \cdots & s_m \\ t_1 & t_2 & \cdots & t_m \end{matrix}\right)}{\Delta\left(\begin{matrix} s_1 & \cdots & s_m \\ t_1 & \cdots & t_m \end{matrix}\right)}, \qquad \Phi_2(s) = \frac{\Delta\left(\begin{matrix} s_1 & s & s_3 & \cdots & s_m \\ t_1 & t_2 & t_3 & \cdots & t_m \end{matrix}\right)}{\Delta\left(\begin{matrix} s_1 & \cdots & s_m \\ t_1 & \cdots & t_m \end{matrix}\right)}, \cdots$$

for which he shows that they are linearly independent and that every other solution of the homogeneous equation is a linear combination of the Φ_j $(1 \leq j \leq m)$. He concludes the theory (which one often calls the "Fredholm alternative") by giving necessary and sufficient conditions on φ for the existence of a solution of (1) when λ is a zero of order m of Δ. He observes that the "transposed equation" obtained from (1) by replacing $K(s,t)$ by $K(t,s)$ has the same "determinant",

and therefore the corresponding homogeneous equation has
exactly m linearly independent solutions $\Psi_1, \ldots, \Psi_m$; the
condition φ must satisfy are then

$$(22) \qquad \int_a^b \varphi(x)\Psi_j(x)dx = 0 \qquad \text{for} \quad 1 \leq j \leq m.$$

Finally, Fredholm shows that for any two kernels K, K′, if
Δ_K and $\Delta_{K'}$ are the corresponding "determinants", then for
the "composed" kernel K″ defined by (17), one has

$$(23) \qquad\qquad \Delta_{K''} = \Delta_K \Delta_{K'}$$

which justifies the name "determinant". He also points out
that his results can be generalized when the kernel K is not
bounded any more, but such that $(x-y)^\alpha K(x,y)$ remains bounded,
with $0 < \alpha < 1$; and he mentions that the extension of his
theorems to any number of variables is immediate.

This beautiful paper may be considered as the source from
which all further developments of spectral theory are derived.
It made a deep and lasting impression on the mathematical
world, and almost overnight the theory of integral equations
became a favorite topic among analysts ([23], [175], [107]).

§2 - The contributions of Hilbert

One of the most active proponents of the new theory was
David Hilbert. As soon as he heard of Fredholm's results, he
started doing himself research work on these questions, made
them one of the main subjects discussed in his Seminar at

Göttingen[*] and supervised many dissertations on the various
aspects of the theory. Between 1904 and 1906, he published
six papers on integral equations in the Göttingen Nachrichten,
later brought together in a single volume entitled "Grundzüge
einer allgemeinen Theorie der Integralgleichungen" [112].

In his first paper [112, p.1-38], Hilbert starts by doing
explicitly what had only been hinted at by Volterra and
Fredholm, the "passage to the limit" in the system (2), res-
tricting himself (as he will do in almost all his results) to
the case in which the kernel K is symmetric, i.e. a real
continuous function such that $K(t,s) = K(s,t)$. He soon realized that in that particular case he might obtain much more
precise results than Fredholm. In the first place, the symmetric matrix $(K(y_k,y_j))$ is then the matrix of the quadratic
form $\sum_{j,k} K(y_k,y_j)\xi_k\xi_j$, and Hilbert undertook to apply also
his "passage to the limit" to that form. He thus obtained
the results which Poincaré had foreseen in the particular
case he had considered (chap.IV, §4): the roots of the Fredholm determinant are then real; if they are written as a sequence (λ_n), each being counted with its multiplicity, then,
for each n there is an eigenfunction φ_n, such that
$\int_a^b \varphi_m(t)\varphi_n(t)dt = 0$ for $m \neq n$. Finally, if one normalizes

(*)
    It is reported (by Hellinger) that Hilbert inaugurated a
session of his Seminar by announcing the development of a
method which would lead to the proof of the Riemann hypothesis:
the problem is to prove that a particular entire function has
all its zeroes on the real line, and Hilbert hoped that this
function would be expressed as the "determinant" of an integral equation with symmetric kernel. However, nobody has yet
been able to find such an equation.

the φ_n by the condition $\int_a^b \varphi_n(t)^2 dt = 1$, and if for each

continuous function x in $[a,b]$, one defines the "Fourier

coefficients" $(x|\varphi_n) = \int_a^b x(t)\varphi_n(t)dt$, Hilbert proves that

$$(24) \qquad \int_a^b \int_a^b K(s,t)x(s)y(t)dsdt = \sum_n \frac{1}{\lambda_n}(x|\varphi_n)(y|\varphi_n)$$

for any two continuous functions x, y, a relation which he

rightly considers as the natural generalization of the clas-

sical reduction of a quadratic form to its "axes". What is

particularly interesting in the way Hilbert considers this

formula is that he shows that the righthand side of (24) is

uniformly convergent when the functions x and y are al-

lowed to vary arbitrarily, subject only to the conditions

$\int_a^b x(t)^2 dt \le 1$ and $\int_a^b y(t)^2 dt \le 1$, the first prefiguration

of what will become "the unit ball in Hilbert space" a few

years later. Of course Hilbert also justifies for his inte-

gral equation the variational definition of the eigenvalues

λ_n first proposed by Weber (chap.III, §1). He shows that the

set of the λ_n is infinite, except when $K(x,y)$ is a linear

combination of a finite number of functions of type $u(x)v(y)$.

He also proves that the resolvent kernel $R(s,t;\mu)$ (in the

sense of Fredholm) has the eigenvalues $\lambda_n-\mu$, the correspond-

ing eigenfunctions being $\varphi_n/(\lambda_n-\mu)$ (μ distinct from the λ_n)

and writes the identity

$$(25) \qquad R(s,t;\mu) - R(s,t;\nu) = (\mu-\nu)\int_a^b R(s,\xi;\mu)R(\xi,t;\nu)d\xi$$

for μ and ν distinct from the λ_n. Finally, he shows that

if a function f can be written in the form

$$(26) \qquad\qquad f(s) \;=\; \int_{a}^{b} K(s,t)g(t)dt$$

for a continuous function   g,   then the corresponding "Fourier
expansion"

$$(27) \qquad\qquad f(s) \;=\; \sum_{n} (f|\varphi_{n})\varphi_{n}(s)$$

is absolutely and uniformly convergent, and one has the
"Parseval identity"

$$(28) \qquad\qquad \int_{a}^{b} f(s)^{2}ds \;=\; \sum_{n} (f|\varphi_{n})^{2}.$$

However, he could only give that proof under the restrictive
assumption that any continuous function could be approximated
(in the sense of mean square value, or, as we would now say,
for the topology of Hilbert space!) by functions of the form
(26).

The proof that this last condition is superfluous was given
in 1905 in the dissertation of Erhard Schmidt, one of the best
students of Hilbert [191]; it contained otherwise no startling
new results, but it deserves some comments, since it is the
first attempt to do away with the Fredholm "determinants", and
substitute to them a more conceptual approach[*].

E. Schmidt begins by proving the Bessel identity

$$(29) \quad \int_{a}^{b} (f(s) \;-\; \sum_{n=1}^{N} (f|\varphi_{n})\varphi_{n}(s))^{2}ds \;=\; \int_{a}^{b} f(s)^{2}ds \;-\; \sum_{n=1}^{N} (f|\varphi_{n})^{2}$$

for an <u>arbitrary</u> orthonormal system (φ_{n}),   from which he

---

[*] Some of the results of E. Schmidt were also obtained independently by W. Stekloff [204].

deduces that for any continuous functions f, g,   the series
$\sum\limits_n (f|\varphi_n)(g|\varphi_n)$ is absolutely convergent, and the convergence
is uniform when f is allowed to vary subject to the condi-
tion $\int_a^b f(s)^2 ds \leq A$ for a fixed constant A.

Next he assumes the existence of the eigenvalues λ_n and of
the corresponding normalized eigenfunctions φ_n, and using
the Bessel inequality, he proves that

$$(30) \qquad \sum_n \frac{1}{\lambda_n^2} \leq \int_a^b \int_a^b K(s,t)^2 \, ds\,dt$$

from which it follows that each λ_n has finite multiplicity
and that $|\lambda_n|$ tends to $+\infty$ with n if there is an infinity
of eigenvalues.

To prepare for the proof of the existence of the eigenvalues,
he introduces, as Fredholm and Volterra had done, the iterated
kernels

$$(31) \qquad K_m(s,t) = \int_a^b K_{m-1}(s,\xi)K(\xi,t)d\xi \text{ for } m > 1, \text{ with } K_1 = K$$

and shows that, if φ is an eigenfunction for K_m, it is
also an eigenfunction for K if m is odd, and is sum of two
eigenfunctions for K if m is even. This allows him to
apply Schwarz's method to prove the existence of at least an
eigenvalue when K is not identically 0, as we have shown
in chapter III, §1, because what he gets in this way is an
eigenvalue of K_2.

Finally, for functions f given by (26), he obtains the
convergence of the Fourier expansion (27) by applying his
initial lemma to the functions $t \mapsto K(s,t)$ and g; and from

that he derives Hilbert's formula (24) by multiplying the
formula $x(s) = \sum_{n} (x|\varphi_n)\varphi_n(s)$ by $K(s,t)y(t)$ and integrat-
ing.

Hilbert's interest in integral equations with symmetric
kernels of course stemmed from the possibility of applying
them to questions of Analysis such as the Dirichlet problem;
it is to such applications that he devoted the second and
third of his papers on integral equations. We shall bypass
them for the time being, as well as most results in his two
last papers on the subject (see chapters VII and IX), to con-
centrate on his fourth paper, published in 1906, a masterpiece
and one of the best papers he ever wrote. By the depth and
novelty of its ideas, it is a turning point in the history of
Functional Analysis, and indeed deserves to be considered as
the very _first_ paper published in that discipline.

Hilbert's new departure in that paper is clear from the
beginning: he deliberately abandons the point of view of integral equations, to return to the older conception of the infinite systems of linear equations (chap.IV, §2), but with a
new twist. This is because he realizes that the theory of
integral equations can be subsumed as a special case of that
older theory: indeed, let (ω_n) be a complete orthonormal
system of continuous functions in $[a,b]$, and suppose the
continuous function f is a solution of (1) for $\lambda = 1$;
then, if we consider the "Fourier coefficients"

$$(32) \quad k_{pq} = \int_a^b \int_a^b K(s,t)\omega_p(s)\omega_q(t)\,ds\,dt,$$

$$b_p = \int_a^b \varphi(s)\omega_p(s)\,ds, \qquad\qquad x_p = \int_a^b f(s)\omega_p(s)\,ds$$

the x_p $(p=1,2,\ldots)$ satisfy the infinite system of linear equations

$$(33) \qquad x_p + \sum_{q=1}^{\infty} k_{pq}x_q = b_p \qquad (p=1,2,\ldots) .$$

The new twist is that, due to the Bessel identity, one has

$$(34) \qquad \sum_{p,q} k_{pq}^2 < +\infty, \qquad \sum_{p} b_p^2 < +\infty, \qquad \sum_{p} x_p^2 < +\infty .$$

Conversely, suppose we have a solution (x_p) of (33) (with conditions (34)), and observe that if $k_q(s) = \int_a^b K(s,t)\psi_q(t)dt$, the functions k_q are continuous, and

$$\sum_{p} k_p(s)^2 \leq \int_a^b K(s,t)^2 dt;$$

the series $u(s) = \sum_{p} x_p k_p(s)$ is then absolutely and uniformly convergent; hence u is continuous and one has $(u|\psi_p) = b_p - x_p$; therefore, if $f = \varphi - u$, $(f|\psi_p) = x_p$ and from the completeness of the system (ψ_p) it follows that f is a solution of (1) for $\lambda = 1$.

Hilbert then embarks into completely uncharted territory:

1º He exclusively considers sequences $x = (x_p)$ (for $p=1,2,\ldots,$) of real numbers such that $\sum_{p} x_p^2 < +\infty$.

2º On the contrary, with regard to the double sequence (k_{pq}) of real numbers, he _abandons_ at first _any_ restrictive condition such as the first condition (34), and only retains the _symmetry_ conditions $k_{qp} = k_{pq}$.

3º The center of interest is not any more the solution of the system (33), but the "symmetric bilinear form"

$$(35) \qquad K(x,y) = \sum_{p,q} k_{pq}x_p y_q$$

which he wants to "reduce" by a formula which would generalize (24).

Of course, even under the restrictions $\sum_p x_p^2 < +\infty$, $\sum_p y_p^2 < +\infty$, the right hand side of (35) is usually meaningless$^{(*)}$; proceeding as Fourier, Poincaré and von Koch (chap.IV, §2), Hilbert considers, for each integer n, the symmetric bilinear form in 2n variables ("sections" (Abschnitte) of K)

$$(36) \qquad K_n(x,y) = \sum_{p=1}^{n} \sum_{q=1}^{n} k_{pq} x_p y_q \, ,$$

but instead of investigating the determinants of these forms, he "reduces" each one to its "axes" and is confronted with the problem of "passing to the limit" for these "reduced" forms when n tends to +∞. We postpone the detailed examination of the original method by which he was able to solve that problem, to chapter VII, which is devoted to the history of modern spectral theory, of which this paper of Hilbert is the starting point; we shall only discuss here the various new notions he is led to introduce in that paper.

A) Hilbert is not yet using the geometrical language which will become prevalent among his immediate successors (cf. §3), but it is obvious that everything he does is inspired by the analogy with n-dimensional Euclidean space. In particular one of his main tools is the generalization of <u>orthogonal trans-</u><u>formations</u>: by that he means that, to every sequence (x_p)

$^{(*)}$To this rather awkward formulation, Hellinger and Toeplitz [106] substituted the consideration of "infinite matrices" $(k_{pq})_{1 \leq p,q < +\infty}$, and of their "calculus" inspired from Frobenius, but without associating an endomorphism to a matrix.

with $\sum\limits_p x_p^2 < +\infty$,   he associates the sequence (x_p'),   where

(37) $x_p' = \sum\limits_q a_{pq} x_q$ $(p=1,2,\ldots)$

and where he imposes on the double sequence (a_{pq}) the conditions

(38)
$$\sum\limits_q a_{pq}^2 = 1   ,   \sum\limits_p a_{pq} a_{pr} = 0   \text{ for } q \neq r$$
$$\sum\limits_p a_{pq}^2 = 1   ,   \sum\limits_q a_{pq} a_{nq} = 0   \text{ for } n \neq p$$

from which he immediately deduces that conversely (x_p) is
deduced from (x_p') by appling the "inverse" orthogonal transformation defined by (a_{pq}') with $a_{pq}' = a_{qp}$.

B) Hilbert restricts himself to forms (35) which he calls
bounded: they are the (not necessarily symmetric) forms such
that there exists an   M > 0   for which one has $|K_n(x,y)| \leq M$
for $\sum\limits_p x_p^2 \leq 1$, $\sum\limits_p y_p^2 \leq 1$ and for all   n;   he also introduces
bounded linear forms $L(x) = \sum\limits_p a_p x_p$,   with $\sum\limits_p a_p^2 < +\infty$,   so
that for any   x   (resp. y),   and any bounded bilinear form K,
the linear forms $K(x,\cdot): x \mapsto K(x,y)$ and $K(\cdot,y): x \mapsto K(x,y)$
are bounded. One of the things he wants to do (inspired of
course by the "reduction" of bilinear forms in a finite number of variables) is to operate an orthogonal transformation
on   x   and   y,   substituting the expressions (37) for the x_p
and doing the same for the y_p.   Unfortunately, he follows
Frobenius in his conception of the "Faltung" of bilinear
forms (instead of the natural idea of "composing" transformations). So, for two bounded bilinear forms A,B, he has to
show that the forms $A_n(x,\cdot) B_n(\cdot,y)$ (Faltung of $A_n(x,y)$
and $B_n(x,y)$, these forms being defined as in (36)) are the

forms $C_n(x,y)$ corresponding to a bounded form $C(x,y)$ which
he calls again the "__Faltung__" of $A(x,y)$ and $B(x,y)$ and
writes $A(x,\cdot)\,B(\cdot,y)$. He can then express the action of an
orthogonal transformation on a bounded bilinear form $K(x,y)$
as a "__Faltung__"

(39) $K'(x',y') = K(\cdot,\cdot)\,0(\cdot,x')\,0(\cdot,y')$

where $0(x,y) = \sum\limits_{p,q} a_{pq}x_p y_q$ is the bounded bilinear form
which he associates to the orthogonal transformation (37).

C) For the development of Functional Analysis, the most important concepts introduced by Hilbert were what he calls
"continuity" and "complete continuity", which correspond to
what will later be called the "strong" and "weak" topologies
on Hilbert space. If $F(x)$ is a complex-valued function defined for all sequences $x = (x_p)$ such that $\sum\limits_p x_p^2 < +\infty$,
Hilbert says that F is __continuous__ if $F(x^{(n)})$ tends to
$F(x)$ when $\sum\limits_p (x_p - x_p^{(n)})^2$ tends to 0, and that F is __com-__
__pletely continuous__ if $F(x^{(n)})$ tends to $F(x)$ when $\sum\limits_p x_p^2 \leq 1$,
$\sum\limits_p (x_p^{(n)})^2 \leq 1$ and each coordinate $x_p^{(n)}$ tends to x_p. He
shows that a bounded bilinear form $K(x,y)$ is continuous, and
that $K_n(x,y)$ tends to $K(x,y)$ when n tends to $+\infty$. But
he pays special attention to the completely continuous symmetric bilinear forms, and gives a separate proof that an orthogonal transformation can reduce any such form to the type

(40) $K(x,y) = \dfrac{1}{\lambda_1} x_1 y_1 + \dfrac{1}{\lambda_2} x_2 y_2 + \dots + \dfrac{1}{\lambda_n} x_n y_n + \dots$

where the sequence $(|\lambda_n|)$ is either finite or tends to $+\infty$.
He realizes that this is a genuine generalization of formula

(24), which is the special case in which $\sum\limits_{p,q} k_{pq}^2 < +\infty$ (corresponding to what will later be called the Hilbert-Schmidt operators); he also mentions another special case, the one in which $K(x,x) > 0$ and $\sum\limits_{p} k_{pp} < +\infty$ (corresponding to the positive nuclear operators of a later date). This formula (40) enables him to go beyond Fredholm by solving a system (33) which is not derived any more from an integral equation, but in which the k_{pq} are only supposed to be such that the symmetric bilinear form (35) is completely continuous. A final remark is that he repeatedly uses with great power what he calls a "principle of choice", which is equivalent to what will later be called the compactness of the unit ball for the weak topology, and that he extends his results to hermitian sesquilinear forms

$$(41) \qquad\qquad K(x,y) = \sum\limits_{p,q} k_{pq} x_p \bar{y}_q$$

where this time the sequences (x_p), (y_p) and (k_{pq}) consist of complex numbers, with $k_{qp} = \bar{k}_{pq}$.

§3 - The confluence of Geometry, Topology and Analysis

It may seem obvious to us that the results of Hilbert are but one step removed from what we now call the theory of Hilbert space; but if, in fact, the birth of that theory almost immediately followed the publication of Hilbert's papers, it seems to me that it is due to the fact that this publication precisely occurred during the emergence of a new concept in mathematics, the concept of structure.

Until the middle of the XIXth century, mathematicians had

been dealing with <u>well determined</u> mathematical "objects":

numbers, points, curves, surfaces, volumes, functions, opera-

tors. But the fact that algebraic manipulations on different

kinds of "objects" had a strikingly similar appearance soon

attracted attention (cf. chap.IV, §3), and after 1840 it

gradually became clear that the essence of these manipulations

did not lie in the <u>nature</u> of the objects, but in the <u>rules</u> to

be followed in handling them, which might be the same for very

different types of objects. However, a precise formulation of

this idea had to wait for the adoption of the set-theoretic

concepts and language; and it is only in 1895 that our defi-

nition of a <u>group</u>, on an <u>arbitrary</u> underlying set, was formu-

lated by Weber [225]. The trend towards the definition of

algebraic structures then gained momentum, and around 1920 all

fundamental notions of present-day Algebra had been defined.

In Analysis, no similar development had yet occurred in

1900. The extensions of the ideas of limit and continuity

which had been formulated always were relative to <u>special</u>

objects such as curves, surfaces or functions. The possibi-

lity of defining such notions in an <u>arbitrary</u> set is an idea

which undoubtedly was first put forward by Fréchet in 1904

[69], and developed by him in his famous thesis of 1906 [71].

The simplest and most fruitful method which he proposed for

such definitions was the introduction of the notion of <u>dis-</u>

<u>tance</u> (which he called "écart") on a set E, a function

$d(x,y)$ defined for any pair (x,y) of elements of E, with

values ≥ 0 and such that: 1) the relation $d(x,y) = 0$ is

equivalent to $x = y$; 2) $d(y,x) = d(x,y)$; 3) $d(x,z) \leq$
$\leq d(x,y) + d(y,z)$ for any three elements of E. It is extremely remarkable that with such simple axioms it is possible
to extend most notions and arguments relative to neighborhoods,
limits and continuity in the space $\mathbb{R}^n$, which usually are introduced in relation to euclidean distance. But the greatest
merit of Fréchet lies in the emphasis he put on three notions
which were to play a fundamental part in all later developments of Functional Analysis: compactness, completeness and
separability. Moreover, he did not limit himself to deriving
general theorems in an abstract setting, but more than half
of his thesis is devoted to very "concrete" metric spaces (as
they came to be called later) closely linked to Analysis: the
space of continuous real functions on a compact interval of $\mathbb{R}$
with the topology of uniform convergence, the space $\mathbb{R}^{\mathbb{N}}$ of
all sequences $n \mapsto x_n$, with the topology of simple convergen-
ce, the space of holomorphic functions in the disc $|z| < 1$,
with the topology of uniform convergence in compact subsets,
and finally the space of all continuous "curves", images of
$[0,1]$ in $\mathbb{R}^3$ by continuous maps, with a "distance" which is
a special case of what was later called the Hausdorff distance between two compact sets.

Clearly Hilbert's work immediately lent itself to application of these ideas, and even invited a bodily transfer of
euclidean geometry in "infinite dimension". This is exactly
what was done by Fréchet himself [72] and by E. Schmidt [192]
in 1908. In E. Schmidt's paper, we find the definition of
what we now call the (complex) space ℓ^2 (or $\ell_{\mathbb{C}}^2$), with the

notions of scalar product and of norm (already written $\|A\|$),
the definition of orthogonality, of closed sets, and of vector
subspaces (called "lineares Funktionengebilde"). The most
interesting feature of that paper is the proof of the existence of the orthogonal projection of a point on a closed vector
subspace, and the purely geometric way in which Schmidt uses
this result to discuss the most general system of linear
equations in Hilbert space

$$(42) \qquad\qquad (x|a_n) = c_n \qquad (n=1,2,\dots)$$

where the a_n are arbitrary vectors of ι^2 and the c_n arbitrary complex numbers.

For each n, Schmidt considers the closed linear affine
varieties F_n of ι^2 defined by the equations $(x|a_j) = c_j$
for $1 \le j \le n$, and the orthogonal projection $x^{(n)}$ of the
origin on F_n; the necessary and sufficient condition of
existence of a solution of the system (42) is that the in-
creasing sequence $(\|x^{(n)}\|)$ be bounded; the sequence $(x^{(n)})$
then has a weak limit x in ι^2, which is the solution of
(42) of smallest norm. Of course, each F_n must be different
from the empty set, which means that any linear relation
$\sum_{k=1}^{n} \lambda_k a_k = 0$ between the vectors a_n must imply $\sum_{k=1}^{n} \lambda_k c_k = 0$;
one can then assume (by dropping some of the equations (42))
that the a_n are linearly independent, and in that case,
Schmidt easily obtains the explicit expression of $\|x^{(n)}\|$:

$$(43) \qquad\qquad \|x^{(n)}\|^2 = \Delta_n/D_n$$

with

$$D_n = \begin{vmatrix} (a_1|a_1) & (a_1|a_2) & \cdots & (a_1|a_n) \\ (a_2|a_1) & (a_2|a_2) & \cdots & (a_2|a_n) \\ \cdots & \cdots & \cdots & \cdots \\ (a_n|a_1) & (a_n|a_2) & \cdots & (a_n|a_n) \end{vmatrix} ,$$

(44)

$$\Delta_n = \begin{vmatrix} 0 & c_1 & c_2 & \cdots & c_n \\ \bar{c}_1 & & & & \\ \vdots & & & D_n & \\ \bar{c}_n & & & & \end{vmatrix}$$

This geometric outlook was already shared in 1906-1907 by two other young mathematicians, E. Fischer and F. Riesz, in the remarkable·work which led them (independently) to what is now called the Fischer-Riesz theorem, introducing a hitherto unsuspected link between Hilbert space and the theory of integration ([66], [183, vol.I, p.378-395]). The latter, from Cauchy to Jordan and Peano, had evolved in a manner completely independent from spectral theory [103]. When Fredholm and E. Schmidt had tried to enlarge the scope of their results on integral equations by weakening the assumptions on the kernel $K(x,y)$, they had nothing else at their disposal beyond the horrible and useless so-called "Riemann integral"[*], and it

---

[*]As a function $K(x,y)$ of two variables may be "Riemann integrable" even if the partial functions $x \mapsto K(x,y)$ are not, Fredholm is compelled to assume that integrability both for the kernel K and all its partial functions! Although E.Schmidt wrote his dissertation in 1905, he probably had no knowledge of Lebesgue's thesis at that time.

is likely that progress in Functional Analysis might have been
appreciably slowed down if the invention of the Lebesgue integral had not appeared, by a happy coincidence, exactly at
the beginning of Hilbert's work on integral equations. With
the help of this marvelous new tool, Fischer and F. Riesz
could define the space $L^2(I)$ over a compact interval $I \subset \mathbb{R}$,
consisting of square integrable functions, when two functions
are identified if they only differ in a set of measure 0.
Their fundamental result is that, if to each function $f \in L^2(I)$,
one associates the sequence (x_p) of its Fourier coefficients
with respect to a complete orthonormal system (equations (32)),
this defines an _isomorphism_ of $L^2(I)$ onto ℓ^2; from that it
follows that $L^2(I)$ is complete and separable. A byproduct
was of course that the results of Fredholm and E. Schmidt
could be applied without change to any integral equation where
the kernel is only supposed to belong to $L^2(I \times I)$, since it
is then equivalent to a system of linear equations corresponding to a "completely continuous" bilinear form in the sense
of Hilbert.

   But the most important consequence of the Fischer-Riesz
theorem is that it opened the way to the definition of the
L^p spaces and to the general theory of normed spaces, which
will be the subject of the next chapter.

CHAPTER VI

DUALITY AND THE DEFINITION OF NORMED SPACES

§1 - The search for continuous linear functionals

In chap.IV, §3, we saw that in 1897 C. Bourlet solved for
the first time the problem of the determination of a linear
map $U: E \to F$ between "function spaces" by conditions of
continuity. In a short Note published in 1903 ([94], vol.I,
p.405-408) Hadamard attacked the same problem with
$E = \mathcal{C}([a,b])$, space of real continuous functions in an interval $[a,b]$, $F = \mathbb{R}$, and "continuity" means for him that
$U(f_n)$ tends to $U(f)$ when f_n tends to f uniformly. He
chooses a fixed function F such that for any continuous
function f, one has

$$(1) \qquad f(x) = \lim_{n \to \infty} n \int_a^b f(t)F(n(t-x))dt$$

uniformly in x; one has then

$$(2) \qquad U(f) = \lim_{n \to \infty} \int_a^b f(t)\Phi_n(t)dt$$

where $\Phi_n(t)$ is the value of U at the function $x \mapsto nF(n(t-x))$;
one may take $F(x) = e^{-x^2}$, so that Φ_n is continuous, but
the choice of F is largely arbitrary (the argument is a
typical case of what later will be called a "regularization"

process).

In two papers published in 1904 and 1905, Fréchet gave
another proof of Hadamard's theorem, and, what is more interesting, began to investigate the similar problems when
$C([a,b])$ is replaced by another "function space"; for instance [70], he remarked that if one takes for E the space
$\mathfrak{B}([a,b])$ of all bounded integrable functions in $[a,b]$ (continuous or not) with the topology of uniform convergence,
there were other continuous linear functionals than those
given by Hadamard's formula, for instance the mappings
$f \mapsto c_1 f(x_1) + \ldots + c_m f(x_m)$, where the x_j are arbitrary
points of $[a,b]$ and the c_j constants. Similarly, if one
takes for E the space of all C^r functions in $[a,b]$, where
convergence means uniform convergence for the function and
its derivatives up to order r, Fréchet showed that the continuous linear functionals could then be written

$$f \mapsto c_o f(a) + c_1 f'(a) + \ldots + c_{r-1} f^{(r-1)}(a) + \lim_{n \to \infty} \int_a^b f^{(r)}(t) \, \Phi_n(t) dt \ .$$

As soon as the study of Hilbert space began (chap.V, §3),
Fréchet [72] and F. Riesz ([183], vol.I, p.386-388) independently showed that continuous "linear functionals" on Hilbert
space $\ell_{\mathbb{R}}^2$ (for the strong topology) could be written unique-
ly as $x \mapsto (x|a)$ for a vector $a \in \ell^2$.

Finally, in 1909, F. Riesz ([183], vol.I, p.400-402) was able
to give a better form to Hadamard's theorem by removing the
arbitrariness of the sequence (Φ_n); his idea was to use the
Stieltjes integral, as Hilbert had done in his work on spectral
theory (see chap.VII, §2): he showed that any continuous linear

functional $: C([a,b]) \to \mathbb{R}$ could be written <u>uniquely</u>

$$(3) \qquad\qquad U: f \longmapsto \int_a^b f(x)d\alpha(x)$$

where α is a function of <u>bounded variation</u> in $[a,b]$, provided one imposed on α the additional conditions of being
continuous on the left and such that $\alpha(a) = 0$. His method
consists in considering, for any $t \in [a,b]$, the function
$f_t \in C([a,b])$ equal to $x-a$ for $a \leq x \leq t$, and to $t-a$ for
$t \leq x \leq b$, and the function $A: t \longmapsto U(f_t)$; he shows that
this function is Lipschitzian, and takes for $-\alpha(t)$ one of
the "derived numbers" of A at the point t; it is then easy
to show that α is a function of bounded variation, and it is
a standard procedure to modify it in such a way that it satisfies the additional conditions mentioned above without
changing U.

Although the contemporaries did not realize the novelty of
F. Riesz's approach, we are justified in seeing in his results
(as he himself did) a radical departure from the conceptions
of linear algebra prevalent in his time:

1º Whereas, even for the space L^2, it was possible, due to
the Fischer-Riesz theorem, to identify the elements of the
space with sequences of numbers, generalizing the dominant
Cayley concept of linear algebra as a theory of "n-tuples",
no such identification was possible for $C([a,b])$, where one
had to work directly on vectors, and not on their "coordinates".

2º Functions of bounded variation may be discontinuous at a
denumerable set of points, and therefore it was impossible to
<u>identify</u> any more the continuous linear functionals on $C([a,b])$

to the elements of that space (again in contrast to what happened in ℓ^2 according to the Riesz-Fréchet theorem).

These features would be still more conspicuous in the theory of L^p and ℓ^p spaces, which F. Riesz began to investigate in 1910 ([183], vol.I, p.403).

§2 – The L^p and ℓ^p spaces

Once the L^2 spaces had been defined, it was a natural generalization to define similarly the function spaces $L^p(I)$ for any interval $I \subset \mathbb{R}$, as the set of all complex valued measurable functions f defined in I and such that $|f|^p$ is integrable, for any $p > 0$ (two functions being identified if they are almost everywhere equal). The study of these spaces was begun by F. Riesz in a fundamental paper ([183], vol.I, p.441-497), second only in importance for the development of Functional Analysis to Hilbert's 1906 paper (chap.V, §2).

Riesz limited himself from the start to the case $p > 1$, in order to be able to use the Hölder and Minkowski inequalities

$$(4) \quad \left| \sum_{k=1}^{n} a_k b_k \right| \leq \left(\sum_{k=1}^{n} |a_k|^p \right)^{1/p} \left(\sum_{k=1}^{n} |b_k|^q \right)^{1/q} \quad \text{for} \quad \frac{1}{p} + \frac{1}{q} = 1$$

$$(5) \quad \left(\sum_{k=1}^{n} |a_k + b_k|^p \right)^{1/p} \leq \left(\sum_{k=1}^{n} |a_k|^p \right)^{1/p} + \left(\sum_{k=1}^{n} |b_k|^p \right)^{1/p}$$

which he first extended to measurable functions, showing that if $f \in L^p(I)$, $g \in L^q(I)$ then fg is integrable and

$$(6) \quad \left| \int_I f(x)g(x)dx \right| \leq \left(\int_I |f(x)|^p dx \right)^{1/p} \left(\int_I |g(x)|^q dx \right)^{1/q},$$

and that if $f \in L^p(I)$, $g \in L^p(I)$,   then $f+g \in L^p(I)$ and

$$(7) \quad \left(\int_I |f(x)+g(x)|^p dx \right)^{1/p} \leq \left(\int_I |f(x)|^p dx \right)^{1/p} + \left(\int_I |g(x)|^p dx \right)^{1/p} .$$

His central theme is the study of infinite systems of linear equations

$$(8) \qquad\qquad \int_I f(x) g_\alpha(x) dx = c_\alpha$$

where the g_α belong to $L^q(I)$ and one looks for a solution $f \in L^p(I)$; this may be considered as the generalization of the problem E. Schmidt had treated in ℓ^2, due to the Fischer-Riesz theorem (chap.V, §3, equations (42)). In order to adapt Schmidt's method to this problem, F. Riesz begins by extending a number of definitions and results from the theory of Hilbert space: strong convergence of a sequence (f_n) of functions of $L^p(I)$ to $f \in L^p(I)$ is defined as meaning that $\int_I |f(x)-f_n(x)|^p dx$ tends to 0. For weak convergence, he first takes as definition that $\int_a^x f_n(t)dt$ tends to $\int_a^x f(t)dt$ for all numbers $x \in I$; and although he proves a little later that this definition is equivalent to the fact that the integrals $\int_I (f(x)-f_n(x))g(x)dx$ tend to 0 for <u>all</u> $g \in L^q(I)$, he essentially uses the first definition to prove the generalization of Hilbert's "principle of choice" (<u>i.e.</u> the weak compactness of the unit ball in $L^p(I)$), which will be one of his main ingredients in the solution of (8). The other ingredient is derived from a result obtained by E. Landau in 1907 [136]; in 1906, Hellinger and Toeplitz had shown that if a sequence (a_n) is such that the series $\sum_n a_n x_n$ is convergent for <u>all</u> sequences (x_n) in ℓ^2, then (a_n) itself belonged to ℓ^2 [106]; Landau proved that, more generally, if $\sum_n a_n x_n$ is convergent for all sequences (x_n) such that $\sum_n |x_n|^p < +\infty$, then $\sum_n |a_n|^q < +\infty$. Approximating functions of L^p by functions having only denumerably many values, F. Riesz deduced from Landau's result that if, for a measurable function g, the product fg is integrable for all functions $f \in L^p$, then necessarily $g \in L^q$.

His solution of (8) then proceeds along the same lines as E. Schmidt; he starts with a finite system (8), for which, using the standard method of analysis (Lagrange multipliers) he proves the existence and uniqueness of a solution $f \in L^p$ for which $\int_I |f(x)|^p dx$ is minimum. The problem is then to find a necessary and sufficient condition on the c_α such that, when one picks from (8) a finite system corresponding to the indices α in an arbitrary finite subset H, the corresponding minima M_H of the integral $\int_I |f(x)|^p dx$ taken for the "minimal" solutions, are <u>uniformly bounded</u> (independently of H); the use of the two ingredients mentioned above then leads to the existence of a solution of (8) by an argument similar to E. Schmidt's. Of course, an explicit expression of M_H (similar to formula (43) of chap.V, §3) is not available here, and the originality of F. Riesz lies in having found a completely different type of condition, namely the existence of a number $M > 0$ such that, for <u>any</u> finite subset H of indices, and <u>any</u> family $(\lambda_\alpha)_{\alpha \in H}$ of scalars, one has the inequality

$$(9) \qquad |\sum_{\alpha \in H} \lambda_\alpha c_\alpha| \le M \cdot \left(\int_I \sum_{\alpha \in H} |\lambda_\alpha g_\alpha(x)|^q dx \right)^{1/q} .$$

F. Riesz in particular applies this condition to the special case in which the g_α are <u>all</u> the functions of $L^q(I)$; (9) is then equivalent to the <u>continuity</u> in L^q of the linear functional L defined by $L(g_\alpha) = c_\alpha$, and he has thus generalized his previous results on ℓ^2 and $C(I)$, proving what we would now express by the statement that <u>the dual of</u> $L^q(I)$ <u>can be identified with</u> $L^p(I)$.

Of course the name "dual" is not yet used by F. Riesz, but he explicitly considers, for a "bounded" linear mapping T of L^p into itself (defined by the condition that $\int_I |T(f)(x)|^p dx$ remains bounded for all f such that $\int_I |f(x)|^p dx \le 1$), the <u>transposed</u> mapping T' defined by the equation

$$(10) \quad \int_I T(f)(x)g(x)dx = \int_I f(x)T'(g)(x)dx \quad \text{for } \underline{all} \ f \in L^p(I).$$

Indeed, for a function $g \in L^q(I)$, this defines (up to a null set) a unique function $T'(g)$, which (by F. Riesz's previous results) also belongs to $L^q(I)$; furthermore, it is easy to show that the mapping T' of L^q into itself is also linear and "bounded". F. Riesz then used this concept to obtain a necessary and sufficient condition for the mapping T to be <u>bijective</u>: he showed that such a condition is the existence of a number $m > 0$ such that <u>both</u> inequalities

$$(11) \quad \begin{aligned} \int_I |T(f)(x)|^p dx &\ge m \cdot \int_I |f(x)|^p dx \\ \int_I |T'(g)(x)|^q dx &\ge m \cdot \int_I |g(x)|^q dx \end{aligned}$$

are satisfied for all $f \in L^p(I)$ and all $g \in L^q(I)$.

F. Riesz had thus given, for the first time, examples of what we now call <u>reflexive</u> Banach spaces not isomorphic to their dual[*]. In his 1913 book on infinite systems of linear equations ([183], vol.II, p.835-1016 and [184]) he treated in a similar way the ℓ^p spaces for $p > 1$ (defined as the set of sequences (x_n) of complex numbers such that $\sum_n |x_n|^p < +\infty$); in addition he stated without proof that for $p \neq 2$, no isomorphism of ℓ^p and L^p existed any more, in contradistinction to the Fischer-Riesz theorem (<u>ibid</u>. Vol.I, p.444-445).

§3 - The birth of normed spaces and the Hahn-Banach theorem

In 1911, F. Riesz combined his methods for the treatment of the system (8) in L^p with the Hadamard-Riesz theorem on linear functionals in $C([a,b])$ in order to study the systems of linear equations

$$(12) \qquad \int_a^b g_\alpha(x)d\xi(x) = c_\alpha$$

where $[a,b]$ is a compact interval in $\mathbb{R}$, the g_α are continuous in $[a,b]$, the c_α are given scalars, and one has to determine a function ξ of bounded variation in $[a,b]$ satisfying the equations (12) for all α . This may be considered as a generalization of a problem which had first been

---

[*]The dual of $L^1(I)$ for a compact interval $I \subset \mathbb{R}$ was shown to be isomorphic to $L^\infty(I)$ by H. Steinhaus [202]; he uses the fact that in that case $L^2(I) \subset L^1(I)$, and therefore a continuous linear functional on $L^1(I)$ is also continuous on $L^2(I)$.

proposed and solved by T. Stieltjes in 1894, the "moment problem": it consists in determining an increasing function ξ
in $[0,+\infty[$ such that

(13) $\int_0^\infty x^n d\xi(x) = c_n \geq 0$ for $n=0,1,2,\ldots$

(the left hand sides are called the "moments" of the function
ξ, a terminology stemming from probability theory) [205];
the same problem was later considered when the interval $[0,+\infty[$
is replaced by $]-\infty,+\infty[$ (the "Hamburger moment problem") or
by a compact interval $[a,b]$ (the "Hausdorff moment problem")
[3].

The solutions to these "moment problems" consist in giving
explicit conditions on the c_n involving existence (or existence and uniqueness) of the function ξ (or rather of the
measure $d\xi$). The condition given by F. Riesz for the existence of a solution ξ of the general system (12) is similar
to condition (9), namely the existence of a number $M > 0$ such
that, for any finite family $(\lambda_\alpha)_{\alpha \in H}$ of scalars, one has

(14) $\left| \sum_{\alpha \in H} \lambda_\alpha c_\alpha \right| \leq M \cdot \sup_x \left| \sum_{\alpha \in H} \lambda_\alpha g_\alpha(x) \right|$;

(he explicitly observed that the right hand side of this inequality is the limit of the right hand side of (9) when q
tends to $+\infty$). His proof is similar to the proof for (8);
he first restricts himself to the case of finite systems (12),
obtains the existence of a "minimal" solution of such a system,
and then, using a "principle of choice" (in our language, the
weak compactness of the unit ball in the space of Stieltjes
measures), he shows that the condition (14) is sufficient for

an arbitrary system (12); his procedure is more complicated
than for (8), because even in the case of a finite system (12),
there is no more uniqueness for the "minimal" solutions ([183],
vol.II, p.798-827).

We now interpret condition (9) in the following way: first,
if $\sum_{\alpha} \lambda_{\alpha} g_{\alpha} = 0$ in $L^q(I)$, then $\sum_{\alpha} \lambda_{\alpha} c_{\alpha} = 0$; this implies
that, if F is the vector subspace of $L^q(I)$ generated by
the g_{α}, there is a well determined linear form L defined
in F such that $L(g_{\alpha}) = c_{\alpha}$ for every α. Condition (9)
then means that this linear form L is <u>continuous</u> in F; the
existence of an $f \in L^p(I)$ such that $L(g_{\alpha}) = \int_I f(x) g_{\alpha}(x) dx$
for all α then means that L can be <u>extended</u> to a continuous
linear form defined in the <u>whole</u> space $L^q(I)$; in other words,
it is a special case of what we now call the Hahn-Banach theorem. There is a similar interpretation of condition (14),
replacing $L^q(I)$ by $C([a,b])$.

Such an interpretation of his results was <u>not</u> given by
F. Riesz; the first mention of that point of view appears in
a paper written in 1912 by the Austrian mathematician E.Helly
(1884-1943), in which he gives a different proof of F. Riesz's
results on the systems (12) [107 bis]. After an interval of
9 years (due to the first World War, in which he was a prisoner
of war in Russia), Helly returned to his method in a paper of
1921 [108] which again should be considered as a landmark in
the history of Functional Analysis, since instead of considering <u>special</u> spaces such as the ℓ^p, L^p or $C([a,b])$, he for
the first time deals with <u>general</u> "normed sequence spaces"
by methods which do not depend on special features of the
space, contrasting with the ones used by E. Schmidt and

F. Riesz.[(*)]

Helly considers vector subspaces of the vector space $\mathbb{C}^{\mathbb{N}}$ of all sequences of complex numbers, and assumes that on such a subspace E there has been defined a <u>norm</u> $\|x\|$ (he does not use that name nor the notation) such that: 1) $\|x\| \geq 0$ and the relation $\|x\| = 0$ is equivalent to $x = 0$; 2) $\|\lambda x\| = |\lambda| \cdot \|x\|$ for any scalar λ; 3) $\|x+y\| \leq \|x\| + \|y\|$; this defines on E a distance $d(x,y) = \|x-y\|$ in the sense of Fréchet. Of course norms had been defined in the spaces ℓ^p, L^p and $\mathbb{C}([a,b])$; but Helly seems to be the first to have noticed the relations of that notion with the concepts of <u>convexity</u> introduced earlier by Minkowski in his "Geometry of numbers" ([161] and [162]). He has shown that the concept of norm on a <u>finite</u> dimensional space $\mathbb{R}^n$ (with the scalars limited to real values) was equivalent to the notion of "symmetric convex body", i.e. a closed, symmetric, bounded convex set in which the origin O is an interior point: such a set B can be defined by an inequality $p(x) \leq 1$, where p is a uniquely determined norm. The boundary of such a set is defined by the equation $p(x) = 1$, and Minkowski had proved that for each point x_o of that boundary, there existed at least one <u>hyperplane of support</u> H, containing x_o and such that B lies entirely <u>on one side</u> of H. If, for an n-tuple of real numbers $u = (u_1,\ldots,u_n)$ and a point $x = (x_1,\ldots,x_n)$,

---

[(*)]It seems that during the period 1910-1920, F. Riesz always had in mind possible axiomatic generalizations of his results, although he did not publish anything in that direction ([183], vol.I, p.452).

one writes $\langle u,x \rangle = u_1 x_1 + \ldots + u_n x_n$, the equation of H has the form $\langle u,x \rangle = 1$ for a suitable u, and one has the inequality $\langle u,x \rangle \leq p(x)$ for all $x \in \mathbb{R}^n$, with $\langle u,x_o \rangle = p(x_o)$; the n-tuples u being identified with the corresponding linear forms $x \mapsto \langle u,x \rangle$ on $\mathbb{R}^n$, Minkowski had also defined the "support function" $q(u) = \sup_{x \neq 0} \langle u,x \rangle / p(x)$, and shown that it was also a norm on $\mathbb{R}^n$, "dual" to p and such that the hyperplanes of support of B are the hyperplanes $\langle u,x \rangle = 1$ with $q(u) = 1$; furthermore, p is the norm "dual" to q, in other words $p(x) = \sup_{u \neq 0} \langle u,x \rangle / q(u)$.

To transfer to spaces of sequences these concepts and definitions, Helly associates to E the subspace E' of $\mathbb{C}^N$ consisting of all the sequences $u = (u_n)$ such that the series $\sum_n u_n x_n$ converges for __all__ $x = (x_n)$ in E [*], and he then considers $\langle u,x \rangle = \sum_n u_n x_n$. For any $u \in E'$, the number $\|u\| = \sup_{x \neq 0} \langle u,x \rangle / \|x\|$ defines a norm on E' provided it is not 0 for some elements $u \neq 0$. Excluding that case, Helly first obtains a weak generalization of Minkowski's result on the hyperplanes of support; if B is the subset of E defined by $\|x\| \leq 1$, he shows that the hyperplane H defined by $\langle u,x \rangle = 1$ meets B if $\|u\| < 1$, does not meet B for $\|u\| > 1$, but if $\|u\| = 1$, the intersection $H \cap B$ may very well be empty: an example is given by taking $E = \ell^1$, $E' = \ell^\infty$ and for H the hyperplane $\sum_{n=1}^{\infty} (1 - \frac{1}{n}) x_n = 1$.

The central problem in Helly's paper is the solution of a system

---

[*] This is not always the dual of E as we now understand that word.

(15) $\langle u^{(\nu)}, x \rangle = c_\nu$ $(\nu=1,2,\ldots)$

where the $u^{(\nu)}$ belong to E' and one looks for a solution $x \in E$. The inequality $|\langle u,x \rangle| \leq \|u\| \cdot \|x\|$ immediately yields the necessary condition similar to Riesz's conditions (9) and (14), namely the existence of a number $M > 0$ such that

(16) $$\left| \sum_{\nu=1}^{n} \lambda_\nu c_\nu \right| \leq M \cdot \left\| \sum_{\nu=1}^{n} \lambda_\nu u^{(\nu)} \right\|$$

for any n and **all** choices of scalars λ_ν; but the example given above (for a single equation) shows that there may well be no solution such that $\|x\| = M$, even when condition (16) is satisfied.

Helly, as Schmidt and F. Riesz had done, first considers the case of a **finite** system (15) of N equations, where as usual the $u^{(\nu)}$ may be supposed to be linearly independent. The mapping $f\colon x \mapsto (\langle u^{(\nu)}, x \rangle)_{1 \leq \nu \leq N}$ of E into $\mathbb{C}^N$ is then surjective; Helly shows that on $\mathbb{C}^N$, $\|y\| = \inf_{f(x)=y} \|x\|$ is a norm (for us it is the natural norm on $E/f^{-1}(0)$ deduced from the norm on E); condition (16) then guarantees the existence of a solution x of (15) such that $\|x\| \leq M_1$, for **any** $M_1 > > M$ (if not necessarily for $M_1 = M$).

The passage from finite systems (15) to the general case is the most original idea of Helly; he splits the problem in two:

A) Given $M_1 > M$, find a linear form $L\colon E' \to \mathbb{C}$, such that $|L(u)| \leq M_1 \cdot \|u\|$ for all $u \in E'$ and such that $L(u^{(\nu)}) = c_\nu$ for all ν.

B) When such a linear form L has been found, find if possible an element $p \in E$ such that $\langle p,u \rangle = L(u)$ for all

$u \in E'$.

To treat problem A), Helly assumes the additional condition that E' is <u>separable</u> as a metric space; he then proves the existence of a solution (a special case of the Hahn-Banach theorem) in the following way. Let $(p^{(\nu)})$ be a sequence of elements of E' which is dense in that space. Helly chooses an increasing sequence $M < M^{(1)} < M^{(2)} < \ldots < M_1$ of numbers, and the main point of his proof consists in showing that there exists a family (γ_ν) of complex numbers such that, for any pair of integers $m \geq 1$, $n \geq 1$, and any pair of families (λ_ν), (μ_ν) of scalars, one has

$$(17) \quad \left| \sum_{\nu=1}^{n} \lambda_\nu c_\nu + \sum_{\nu=1}^{m} \mu_\nu \gamma_\nu \right| \leq M^{(m)} \cdot \left\| \sum_{\nu=1}^{n} \lambda_\nu u^{(\nu)} + \sum_{\nu=1}^{m} \mu_\nu p^{(\nu)} \right\|.$$

It is then easy to show that there exists a linear form L on E' such that $L(p^{(\nu)}) = \gamma_\nu$ for all indices ν, and that it is a solution of problem A).

The proof of (17) is done by induction on m, the case $m = 0$ being the assumption (16). One has then to prove the existence of a point $\gamma_{m+1} \in \mathbb{C}$ which, for any integer $n \geq 1$ and any pair of families of scalars $(\lambda_\nu)_{1 \leq \nu \leq n}$, $(\mu_\nu)_{1 \leq \nu \leq m}$, belongs to <u>all</u> disks defined in $\mathbb{C}$ by

$$(18) \quad \left| \sum_{\nu=1}^{n} \lambda_\nu c_\nu + \sum_{\nu=1}^{m} \mu_\nu \gamma_\nu + \gamma_{m+1} \right| \leq$$
$$\leq M^{(m+1)} \left\| \sum_{\nu=1}^{n} \lambda_\nu u^{(\nu)} + \sum_{\nu=1}^{m} \mu_\nu p^{(\nu)} + p^{(m+1)} \right\|.$$

However, a general result on convex sets in a finite dimensional space, proved by Helly himself, reduces that question to proving that <u>any three</u> of the disks (18) have a common point;

and this is shown by Helly to be a consequence of the result
proved before for <u>finite</u> systems (15).

Turning to problem B), Helly discovers that it is quite possible that it has <u>no</u> solution; in our language, he gives the
first example of <u>non reflexive</u> Banach spaces$^{(*)}$. That example
is the **space** E of all sequences (x_k) such that the series
$\sum\limits_{k=1}^{\infty} x_k$ converges, with the norm $\|x\| = \sup\limits_{n}\left|\sum\limits_{k=n}^{\infty} x_k\right|$; Helly
proves that E' consists of all sequences (u_k) such that
$\|u\| = |u_1| + \sum\limits_{k=1}^{\infty} |u_{k+1}-u_k|$ is finite, $\|u\|$ being the natural
norm on E'; then if one takes $L(u) = \lim\limits_{k\to\infty} u_k$, L is continuous on E' but there is no $p \in E$ such that $L(u) = \langle p,u \rangle$.

Starting from the work of F. Riesz and Helly, it was a natural generalization to define norms on <u>arbitrary</u> vector
spaces over ℝ or ℂ, and not only on spaces of functions
or on subspaces of $ℂ^N$. This was done independently by
H. Hahn [97] and S. Banach [12], who restrict themselves to
<u>complete</u> spaces.

Banach's paper is his thesis, written in 1920: although he
does not mention convexity, he is careful to develop and extensively use a geometric language. He is mainly interested
in continuous linear operators u: E → F, where E and F
are arbitrary normed complete spaces, and in limits of sequences of such operators. Hahn's point of view is similar,

---

$^{(*)}$F. Riesz had already observed that one could define on the
space of functions of bounded variation continuous linear
funcionals which were not of the form $\xi \mapsto \int_a^b f(x)d\xi(s)$ for a
continuous function f (for instance one can take for f an
increasing discontinuous function) ([183], vol.II, p.827).

although he is only concerned with linear forms; neither he
nor Banach are at that moment interested in the problem of
extension of linear forms, and we postpone a more detailed
discussion of their papers of 1922-23 to §4. We should however mention that in his thesis Banach gives the "abstract"
formulation of the method of successive approximations (chap.
II, §1) as a "contraction principle": if F is a mapping of
a complete normed space E into itself such that $\|F(x)-F(y)\|$
$\leq k\|x-y\|$ with $0 < k < 1$, then the sequence (x_n) defined
by induction as $x_{n+1} = F(x_n)$ $(x_o$ arbitrary) converges to
the unique "fixed point" x, such that $F(x) = x$.

It was only in 1927 that Hahn returned to Helly's paper, in
the general context of complete normed spaces, and completely
solved the extension problem for such spaces [98]. He proceeds
by induction as Helly had done, but at the same time he greatly simplifies and generalizes the method by introducing, for
the first time in general problems of Functional Analysis [(*)],
transfinite induction instead of the ordinary kind. In a complete normed space E, one has a vector subspace V and
there is defined on V a (real valued) linear form f such
that $|f(x)| \leq M\|x\|$ for $x \in V$; the problem is to extend f
to a linear form F on E such that $|F(y)| \leq M\|y\|$ for
$y \in E$. Hahn begins by showing the existence of an ordinal γ,
and of a mapping $\xi \mapsto V_\xi$ which, to very ordinal $\xi < \gamma$ asso-

[(*)]Transfinite induction had been used by analysts ever since
Cantor, but the application of transfinite induction closest
to Hahn's is probably the method by which Banach, in 1923,
had proved the existence on $\mathbb{R}$ of a "measure" defined on all
subsets of $\mathbb{R}$ and simply additive [13].

ciates a vector subspace V_ξ of E such that $V_o = V$, $V_\xi \subset V_\eta$ for $\xi < \eta$, V_ξ has codimension 1 in $V_{\xi+1}$ and E is the union of the V_ξ for $\xi < \gamma$. The problem is then easily reduced to the case in which V has codimension 1 in E, and then E is generated by V and an element $a \notin V$; Hahn considers the l.u.b B of the numbers $f(x) - M\|x-a\|$ for $x \in V$, and the g.l.b. A of the numbers $f(x) + M\|x-a\|$ for $x \in V$, and, using the assumption $|f(x)| \le M\|x\|$ for $x \in V$, he easily shows that $A \le B$; the extension F is then defined by $F(x+\lambda a) = f(x) + \lambda c$ for all $\lambda \in \mathbb{R}$, where c is any number such that $A \le c \le B$.

As a particular case of his theorem, Hahn shows that for any vector $a \ne 0$ in E, there exists a continuous linear form L on E such that $\|L\| = 1$ and $L(a) = \|a\|$; he then formally introduces the <u>dual</u> space E' of E ("polare Raum" in his terminology) which is not reduced to 0 due to the preceding result; he writes $B(u,x)$ instead of $u(x)$ for $x \in E$, $u \in E'$, and considers for any $x \in E$, the linear form $c(x): u \mapsto B(u,x)$ on E', for which he shows that $\|c(x)\| = \|x\|$. In other words, he has defined a linear isometry c of E into its second dual E'', and he says a space E is "regular" if c is bijective (our reflexive spaces). It may therefore rightly be said that with this paper of Hahn, duality theory at last had come into its own.

Two years later, Banach, who apparently was not aware of Hahn's paper, published the same theorem with the same proof (he later acknowledged Hahn's priority); in addition, he recognized that the argument could be generalized: if p is a real valued function defined in a vector space   E   and such

that $p(x+y) \leq p(x) + p(y)$ and $p(\lambda x) = \lambda p(x)$ for $\lambda \geq 0$,

and if   f   is a linear form defined in a vector subspace   V

of   E   and such that $f(x) \leq p(x)$ in   V,   then it is possible

to extend   f   to a linear form   F   defined in   E   and such

that $F(x) \leq p(x)$ in   E.   This extension was to play later

an important role in the development of the theory of locally

convex spaces (cf. chapter VIII).

§4 - The method of the gliding hump and Baire category

In his 1922 paper [97], Hahn proved the following theorem:

let   E   be a complete normed space, (u_n) a sequence of con-

tinuous linear forms on   E,   and suppose that for each $x \in E$,

the sequence of numbers $|u_n(x)|$ is bounded by a number <u>de-

pending on x</u>; then the sequence of the norms $\|u_n\|$ is <u>bounded</u>.

The proof is by contradiction; assuming that the sequence

$(\|u_n\|)$ is unbounded, one determines by induction a sequence

(x_k) in   E   and a sequence (n_k) of integers such that:

1º   the series $\displaystyle\sum_{k=1}^{\infty} x_k$ converges to an element $x \in E$;

2º $\displaystyle\sum_{j=k+1}^{\infty} |u_{n_k}(x_j)| \leq 1$;

3º $|u_{n_k}(x_k)| \geq k + \displaystyle\sum_{j=1}^{k-1} |u_{n_k}(x_j)|$.

Then one has for each   k,

$$|u_{n_k}(x)| \geq |u_{n_k}(x_k)| - \sum_{j=1}^{k-1} |u_{n_k}(x_j)| - \sum_{j=k+1}^{\infty} |u_{n_k}(x_j)| \geq k-1$$

which contradicts the assumption.   To do this, one assumes

the u_{n_j} have been determined for $j < k$,   and one considers

a ball

$$B_k: \ \|x\| \ \leq \ 2^{-k} \cdot \inf_{j<k} \ (\|u_{n_j}\|+1)^{-1}$$

in E; the assumption that $(\|u_n\|)$ is unbounded guarantees the existence of an index n_k and a point $x_k \in B_k$ for which condition 3º holds; conditions 1º and 2º are then deduced from the choice of the radius of the ball B_k.

This is often called the "method of the gliding hump": in the sequence of values $|u_{n_k}(x_j)|$ when j varies from 1 to $+\infty$, the index $j = k$ corresponds to a "hump" much bigger than the sum of the contributions of the other indices.

The result can be put in a different form: if the sequence $(\|u_n\|)$ is unbounded, there exists at least one $x \in E$ such that the sequence $(|u_n(x)|)$ is unbounded.

In this form, the first example of the method of the gliding hump is probably the way in which Lebesgue, in 1905 ([138], vol.III, p.101] and [139, p.86-88]) constructed a continuous periodic function $F(x)$ in $[0,2\pi]$ whose Fourier series diverges at the point 0. He had proved that, if one writes $S_n(g)$ for the sum of the first n terms of the Fourier series of a continuous function g, it is possible to find a sequence (g_n) of continuous periodc functions of bounded variation such that $|g_n(t)| \leq 1$ in $[0,2\pi]$ and that the sequence of values $S_n(g_n)(0)$ tends to $+\infty$. He then defines

$$F(x) = \varepsilon_1 f_1(n_1 x) + \varepsilon_2 f_2(n_2 x) + \ldots + \varepsilon_k f_k(n_k x) + \ldots$$

where the ε_k are > 0 and such that $\sum_{k=1}^{\infty} \varepsilon_k = 1$, the f_k are continuous periodic functions of bounded variation such that $|f_k(t)| \leq 1$ in $[0,2\pi]$ and that $|S_{p_k}(f_k)(0)| \geq k/\varepsilon_k$ for an increasing sequence (p_k) of integers. Finally the

increasing sequence (n_k) of integers is chosen in such a
way that $n_k > n_{k-1}p_{k-1}$ and that, for the continuous function
of bounded variation $F_k(x) = \varepsilon_1 f_1(n_1 x) + \ldots + \varepsilon_{k-1} f_{k-1}(n_{k-1} x)$,
all the sums $S_n(F_k)(0)$ are ≤ 2 in absolute value for $n \geq n_k$
(they converge to $F_k(0)$). This choice implies that, for
$j > k$, the sum of the first $n_k p_k$ terms of the Fourier
series of $f_j(n_j x)$ is reduced to the <u>first</u> term of the series,
hence is ≤ 1 in absolute value; using these definitions it is
easy to check that $|S_{n_k p_k}(F)(0)| \geq k-3$ for all k.

One year later, Hellinger and Toeplitz, two students of
Hilbert, found a rather surprising complement to the definition he had given of a <u>bounded</u> bilinear form (chap.V, §2);
instead of assuming that $|K_n(x,y)| \leq M$ for all n and all
$x = (x_p)$ and $y = (y_p)$ such that $\sum_p x_p^2 \leq 1$ and $\sum_p y_p^2 \leq 1$,
they showed that it was enough to assume that for each such
pair (x,y), one had $|K_n(x,y)| \leq M_{x,y}$ for all n, where
the number $M_{x,y}$ <u>might depend on</u> x, y <u>in an arbitrary way</u>.
Independently of Lebesgue, they proved that result by a
"gliding hump" method, constructing a pair (x,y) for which
the sequence $(|K_n(x,y)|)$ is unbounded if K is not a bound-
ed form in Hilbert's sense [106].

During the next 20 years, many more examples of the "gliding
hump" method appeared in the literature: Lebesgue used it
repeatedly in a 1909 paper on "singular integrals" ([138],
vol.III, p.259-351), where one looks for conditions on
"kernels" K_n insuring that the integrals $\int_a^b f(t)K_n(t,x)dt$ tend
to $f(x)$ when n tends to $+\infty$, for various kinds of function f.
The method was also prominent in the study of "summation pro-

cesses", where one "transforms" a sequence (x_n) into a se-
quence (y_n) by the formulas $y_n = \sum\limits_{m=1}^{\infty} a_{nm} x_m$, and has to
look for conditions on the a_{nm} insuring that when (x_n) has
a limit, (y_n) tends to the same limit ([193], vol.II,
p. 389-321). Hahn's paper of 1922 [97] was written to give a
general background to all these results, showing that they all
were consequences of his general theorem. Independently,
Banach, in his thesis, proved a theorem more general than
Hahn's, the u_n being now continuous linear operators from a
complete normed space E into a complete normed space F; he
showed that the assumption that the norms $\|u_n(x)\|$ are bound-
ed for each x by a number depending on x, implies that the
sequence of the norms $\|u_n\|$ is bounded.

Finally, in 1927, Banach and Steinhaus (using an idea of
Saks) discovered that this theorem could be proved without
using the "gliding hump" method, by an application of a theorem Baire had proved in 1899 [11]: he had shown that in $\mathbb{R}^n$,
the intersection of a denumerable family of dense open subsets
is itself dense[(*)]; this implies that if u is a real func-
tion defined and <u>lower semi-continuous</u> in $\mathbb{R}^n$, and if $u(x) <$
$< +\infty$ for each $x \in \mathbb{R}^n$, then any non empty open subset U of
$\mathbb{R}^n$ contains a non empty open subset V such that $\sup\limits_{x \in V} u(x) <$
$< +\infty$. These results and their proofs immediately generalize
when $\mathbb{R}^n$ is replaced by an arbitrary <u>complete metric space.</u>
If now H is a set of linear mappings from a complete normed
space E into a complete normed space F, and if for each

[(*)] For $n = 1$, the same result had been proved two years
earlier by W. Osgood [170].

$x \in E$, $\sup\limits_{u \in H} \|u(x)\| < +\infty$, the function $p(x) = \sup\limits_{u \in H} \|u(x)\|$ is

lower semi-continuous, and from the Baire theorem it follows

that p is bounded in a neighborhood of 0, which implies

that $\sup\limits_{u \in H} \|u\|$ is finite [16].

§5 - Banach's book and beyond

In 1932 S. Banach published a book [15] containing a comprehensive account of all results known at that time in the
theory of normed spaces, and in particular the theorems he
had published in his papers of 1923 and 1929. A large part
was devoted to the concept of weak convergence and its generalizations, which he had begun to study in 1929; we shall
postpone to chap.VIII, §1 the discussion of these questions.
The most remarkable result contained in that book is another
consequence of Baire's theorem, discovered by Banach, and much
deeper than the Banach-Steinhaus theorem: if u is a continuous linear mapping from a complete normed space E into a
complete normed space F, then either u(E) is meager in F
(a set "of first category" in the terminology of Baire), or
u(E) = F. An immediate consequence is the famous closed graph
theorem: if u is a linear mapping from E to F having a
closed graph in ExF, then u is continuous. These surprising results have become two of the most powerful tools in all
applications of Functional Analysis.

These features, as well as many applications to classical
Analysis, gave the book a great appeal, and it had on Functional Analysis the same impact that van der Waerden's book

had on Algebra two years earlier. Analysts all over the
world began to realize the power of the new methods and to
apply them to a great variety of problems; Banach's terminology and notations were universally adopted, complete normed
spaces became known as Banach spaces, and soon their theory
was considered as a compulsory part in most curricula of
graduate students. After 1935, the theory of normed spaces
became part of the more general theory of locally convex
spaces, which we shall discuss in chapter VIII; more recently
however, there has been a renewed surge of interest in the
special properties of normed spaces and their "geometry"; it
is too soon, as yet, to have a clear idea of the scope of
these results and of their relation to other parts of mathematics, and we refer the interested reader to [4], [17], [47],
[50], [116], [134], [149], [150] and [185].

CHAPTER VII

SPECTRAL THEORY AFTER 1900

§1 - F. Riesz's theory of compact operators

We already mentioned that Fredholm's paper attracted many
mathematicians to the theory of integral equations, and also
to the theory of infinite systems of linear equations, espe-
cially after Hilbert had given it a new impetus. We shall not
examine these papers, most of which are concerned with special
problems, without much bearing on the progress of Functional
Analysis, and we refer the interested reader to [107] (in particular p.1543-1552 and p.1574-1575).

The "Fredholm alternative" corresponded, in "infinite dimensional linear algebra" to the classical relation between kernel
and image of an endomorphism of a finite dimensional vector
space over $\mathbb{C}$. But for such endomorphisms, much more was
known, namely the Jordan normal form which characterized them
up to "similitude", and a natural question was to investigate
similar properties of the Fredholm operators. However, only
partial results in that direction were obtained, before F.Riesz
in 1916 (in a paper written in Hungarian ([183], vol.II,
p.1017-1052) and only published in German in 1918 (_ibid._,
p.1053-1080)) gave a complete answer to that question, and

found the proper context to Fredholm's results, in what is
now known as the Riesz-Fredholm theory of compact operators.

F. Riesz never adopted Hilbert's method of dealing with linear equations via bilinear forms, but followed Fredholm in
using instead operators. In his work on ℓ^p spaces ([183],
vol.II, p 876-911 and [184]), he therefore had translated
Hilbert's conception of a completely continuous bilinear form
(chap.V, §2) into the notion of completely continuous operator:
for him it was a linear mapping of ℓ^p into itself which
transformed weakly convergent sequences into strongly convergent ones. The novelty in his 1918 paper is that he realized
that he could give an equivalent definition without mentioning
weak convergence, using instead the general concept of compactness introduced by Fréchet: the condition was that the linear
operator transformed a bounded set into a relatively compact
one (for the strong topology). Now this can be defined for an
arbitrary normed space instead of ℓ^p; in his 1918 paper
F. Riesz restricted himself to the space $C(I)$ for a compact
interval $I \subset \mathbb{R}$, but he explicitly mentioned that he merely
considered that case as a "touchstone" for more general conceptions (ibid.,p.1053). And indeed, after he has defined the
norm on $C(I)$, he never (except when proving that the Fredholm operator for continuous kernels is completely continuous
in $C(I)$) uses anything except the axiomatic definition of
a norm (remember that this definition only appeared in print
4 years later!).

In my opinion, F. Riesz's 1918 paper is one of the most
beautiful ever written; it is entirely geometric in language

and spirit, and so perfectly adapted to its goal that it has
never been superseded and that Riesz's proofs can still be
transcribed almost <u>verbatim</u>. He starts from two almost obvious
remarks: 1) in a normed space E, if V is a closed vector
subspace not equal to   E,   there is a vector $x \in E$ such
that $\|x\| = 1$ and $\|x-y\| \geq \frac{1}{2}$ for all $y \in V$; 2) a subset
$S \subset E$ cannot be relatively compact if there is in S an in-
finite sequence (x_n) such that $\|x_j - x_k\| \geq \frac{1}{2}$ for all pairs
of distinct indices. The first consequence is the celebrated
theorem characterizing finite dimensional normed spaces as the
only <u>locally compact</u> ones: one has only to cover the ball
$\|x\| \leq 1$ with a finite number of balls $\|x-a_j\| \leq 1/4$, and
then there cannot be any point such that $\|z\| = 1$ and
$\|z-y\| \geq 1/2$ for all points of the (necessarily closed) vector
subspace V generated by the a_j.

   F. Riesz then considers a completely continuous linear mapping u of E into itself (or, as we now say, a <u>compact</u> linear mapping), and studies the endomorphism $v = 1_E - u$ of E.
Using the two remarks above and very simple arguments, he
proves in succession the following properties:

   a) the kernel $v^{-1}(0)$ has finite dimension;
   b) the image $v(E)$ is closed in E;
   c) the codimension of $v(E)$ in E is finite.

The next step is to consider the <u>iterates</u> v^k of v, the
kernel N_k and the image F_k of v^k (*); the N_k form an

_____

(*)In finite dimensional spaces, this method to obtain
the Jordan normal form of an endomorphism had been developed
by E. Weyr [228].

increasing sequence of closed subspaces of finite dimension, the F_k a decreasing sequence of closed subspaces of finite codimension. F. Riesz shows, by contradiction and using remark 2) above, that there is a smallest integer n such that $N_{k+1} = N_k$ for $k \geq n$; it is then an easy matter to prove that $F_{k+1} = F_k$ for $k \geq n$, and that E is the <u>topological direct sum</u> of F_n and N_n; the restriction of v to F_n is a linear <u>homeomorphism</u> of F_n onto itself. In particular, if $N_1 = v^{-1}(0) = \{0\}$, v is a linear homeomorphism of E onto itself, and its inverse $w = v^{-1}$ is such that $(1_E - u)w = 1_E$, in other words $w = 1_E + uw$ has the same form as v, since uw is compact.

These results enable F. Riesz to treat completely the question of <u>eigenvalues</u> of a compact operator. There are at most denumerably many eigenvalues $\lambda_n \neq 0$ in $\mathbb{C}$, and each of them is <u>isolated</u> in $\mathbb{C}-\{0\}$; their set is bounded and 0 belongs to its closure if it is infinite. For each $\lambda_n \neq 0$, E splits into a topological direct sum of two closed subspaces $F(\lambda_n)$ and $N(\lambda_n)$, which are stable by u; $N(\lambda_n)$ has finite dimension, and there is a smallest integer k_n such that the restriction of $(u-\lambda_n \cdot 1_E)^{k_n}$ to $N(\lambda_n)$ is 0; the restriction of $u-\lambda_n \cdot 1_E$ to $F(\lambda_n)$ is a linear homeomorphism of that subspace onto itself. If E is complete, the function $\zeta \longmapsto (u-\zeta \cdot 1_E)^{-1}$ is meromorphic in $\mathbb{C}-\{0\}$ (with values in the space $\mathcal{L}(E)$ of continuous endomorphisms of E); at the points other than the λ_n, that function is holomorphic, and at each λ_n it has a pole of order k_n. Finally, if $m \neq n$, the subspace $N(\lambda_n)$ is contained in $F(\lambda_m)$. However,

there is in general no global decomposition of E into a sum
of subspaces $N(\lambda_n)$, similar to what happens for compact
self-adjoint operators in Hilbert space [99]. As a matter of
fact, there may be no eigenvalues at all, as for instance for
Volterra operators.

In the study of $u-\zeta \cdot 1_E$, the value $\zeta = 0$ is completely
exceptional; $u(E)$ is not closed in general and may have infinite codimension, and $u^{-1}(0)$ may have infinite dimension.
This explains the intractability of integral equations "of
the first kind", special cases of equations $u(x) = y$ for u
compact, which had baffled the early mathematicians working
on integral equations.

Although there has been much work done on compact operators
of special types, the general theory of compact operators has
remained pretty much what it was after the publication of
F. Riesz's 1918 paper. Among more recent results, one can
mention the fact that when u is a compact operator in a complete normed space, its transposed operator $^t u$ in the dual
E' is also a compact operator [188]. It has also been proved
that, even when a compact operator u in E has no eigenvalue, there are always closed vector subspaces V of E,
different from E and $\{0\}$, such that $u(V) \subset V$ [7].

§2 - The spectral theory of Hilbert

We now return to the most original part of Hilbert's 1906
paper (chap.V, §2), in which he discovered the entirely new
phenomenon of the "continuous spectrum". In his "Theory of
heat", Fourier had considered trigonometric series represent-

ing functions of period   2a   (chap.I, §2, formula (13)) when

a <u>tends to</u> +∞. The eigenvalues $\lambda_n = \left(n+\frac{1}{2}\right)^2 \pi^2/a^2$ of the

corresponding Sturm-Liouville problem for the equation $y'' +\lambda y =$

$= 0$ with boundary conditions $y(-a) = y(a) = 0$ divide the

interval $[0,+\infty[$ in intervals of length tending to 0 with

$1/a$, and this had led Fourier to consider that the "limiting

case" of the trigonometric expansion of a function of period

2a would be, for any function f defined on $\mathbb{R}$, the repre-

sentation by an <u>integral</u>

(1) $$f(x) = \frac{1}{\pi} \int_{-\infty}^{+\infty} dt \int_{0}^{\infty} f(t)\cos u(x-t)du$$

where the "eigenvalues" would now <u>fill the interval</u> $[0,+\infty[$

([67], vol.I, p.392).

In 1897, Wirtinger [230] developed similar ideas for Hill's

equation

(2) $$y'' + \lambda q(x)y = 0$$

where q is a continuous <u>periodic</u> function of period 1. The

general theory of these equations was well-known at the time:

starting with a fundamental system of solutions u_1, u_2 such

that

$$u_1(0,\lambda) = 1, \qquad u_1'(0,\lambda) = 0$$
$$u_2(0,\lambda) = 0, \qquad u_2'(0,\lambda) = -1$$

so that $u_2u_1' - u_1u_2'$ was the constant function 1, one con-

siders complex solutions f such that $f(x+1) = \rho f(x)$ for

all $x \in \mathbb{R}$; the constants $\rho(\lambda)$ having that property are so-

lutions of the equation

(3) $$\rho^2 + \Phi(\lambda)\rho + 1 = 0$$

with $\Phi(\lambda) = u_2'(1,\lambda) - u_1(1,\lambda)$. The solutions of period 1
correspond to "eigenvalues" λ such that $\Phi(\lambda) = 2$, the so-
lutions f such that $|f(x+1)| = |f(x)|$ to values of λ
such that $-2 \leq \Phi(\lambda) \leq 2$, which in general constitute disj-
oint intervals I_k of $\mathbb{R}$. Wirtinger looks for solutions of
period n (an arbitrary integer), which correspond to values
of λ such that $(\rho(\lambda))^n = 1$, and he shows that when n
tends to $+\infty$, these values of λ tend to "fill up" the intervals I_k. The similarity with the optical spectra of molecules leads him to speak of the "Bandesspectrum" of equation (2) formed by the union of the intervals I_k, and he
thinks there should be an integral formula similar to (1),
without being able to guess what that formula could be.

Although Hilbert does not mention Wirtinger's paper, it is
probable that he had read it (it is quoted by several of his
pupils), and it may be that the name "Spectrum" which he used
came from it; but it is a far cry from the vague ideas of
Wirtinger to the extremely general and precise results of
Hilbert. On the other hand, the influence of Stieltjes's big
paper of 1894 on continued fractions is explicitly acknowledged by Hilbert: Stieltjes had had to take the limit of a sequence of rational functions of a complex variable

$$\frac{P_{2n+1}(z)}{Q_{2n+1}(z)} = \sum_{i=1}^{n} \frac{M_i}{z+x_i}$$

where the M_i and the x_i are real numbers, and he had shown
that the limit could be written as a "Stieltjes transform"

$$F(z) = \int_0^\infty \frac{d\Phi(u)}{z+u}$$

for a function Φ of bounded variation (creating for that purpose the concept of "Stieltjes integral") [205]. It is a similar problem which confronts Hilbert when he wants to pass from the classical "reduction" of the n-th "Abschnitt"

$$(4) \qquad K_n(x,x) = \sum_{p=1}^{n} \sum_{q=1}^{n} k_{pq} x_p x_q$$

of his "bounded" quadratic form $K(x,x)$, to a "reduction" of $K(x,x)$ itself: the classical theory shows that one has

$$(5) \qquad K_n(x,x) = \frac{(L_1^{(n)}(x))^2}{\lambda_1^{(n)}} + \dots + \frac{(L_n^{(n)}(x))^2}{\lambda_n^{(n)}}$$

where the $\lambda_j^{(n)}$ are real numbers such that $\lambda_1^{(n)} \le \lambda_2^{(n)} \le \dots \le \lambda_n^{(n)}$, and the $L_j^{(n)}(x)$ are linear forms in $x_1, x_2, \dots, x_n$ such that

$$(6) \qquad (L_1^{(n)}(x))^2 + \dots + (L_n^{(n)}(x))^2 = x_1^2 + \dots + x_n^2 \; .$$

To each $x = (x_p)$ of ℓ^2 such that $x_p = 0$ for $p > n$, Hilbert associates the piecewise linear convex function of the real variable λ

$$(7) \qquad u^{(n)}(x;\lambda) = \sum_{p=1}^{n} (L_p^{(n)}(x))^2 (\lambda - \lambda_p^{(n)})^+ \; .$$

His idea is to take (if possible) the limit of each of these functions, for the points of ℓ^2 having only a finite number of coordinates $\neq 0$. Using his "principle of choice" (chap.V, §2) and Cantor's diagonal process, he shows that these limits exist at least for a suitable subsequence of the $u^{(n)}$. In fact, he only uses these limits for the points $x^{(pp)}$ having the coordinate of index p equal to 1, all others to 0, and the points $x^{(pq)}$ for $p \neq q$, having the coordinates of in-

dices p and q equal to $\dfrac{1}{\sqrt{2}}$ and all others to 0; he
writes the corresponding limits $u_{pq}(\lambda)$ for all pairs of integers (p,q). They are convex functions of λ, hence the
right derivative $v_{pq}^{+}(\lambda)$ and the left derivative $v_{pq}^{-}(\lambda)$
exist for all λ, and are equal except for an at most denu-
merable set of values of λ. Hilbert writes $\lambda_1, \lambda_2, \ldots, \lambda_r, \ldots$
the sequence of these exceptional values of λ for **all** pairs
(p,q), and defines quadratic forms

$$v^{+}(x;\lambda) = \underset{p,q}{\Sigma} \; v_{pq}^{+}(\lambda) x_p x_q, \qquad v^{-}(x;\lambda) = \underset{p,q}{\Sigma} \; v_{pq}^{-}(\lambda) x_p x_q$$

and for each index r,

$$E_r(x) = v^{+}(x;\lambda_r) - v^{-}(x;\lambda_r).$$

All these are bounded quadratic forms, and more precisely,
their values in ℓ^2 are ≥ 0 and $\leq (x|x)$. For the values of
λ distinct from the λ_r, he writes

$$e(x;\lambda) = \underset{\lambda_r < \lambda}{\Sigma} \; E_r(x)$$

and

$$\sigma(x;\lambda) = v(x;\lambda) - e(x;\lambda)$$

where v is the common value of v^{+} and v^{-}. With these
notations, his final result is the "reduction" of the quadratic form $K(x,x)$: for each $x \in \ell^2$, the function
$\lambda \mapsto \sigma(x;\lambda)$ is a continuous function of bounded variation, and
one may write

$$(8) \qquad \begin{cases} (x|x) = \underset{r}{\Sigma} \; E_r(x) + \displaystyle\int d\sigma(x;\lambda) \\[2ex] K(x,x) = \underset{r}{\Sigma} \dfrac{1}{\lambda_r} E_r(x) + \displaystyle\int \dfrac{1}{\lambda} d\sigma(x;\lambda). \end{cases}$$

He says that the set of the λ_r is the <u>point spectrum</u> of K
and the complement of the set where all the $\sigma(x;\lambda)$ are constant the <u>continuous spectrum</u> of K .

Of course, when K is a "completely continuous" form, the
continuous spectrum is absent, but Hilbert gives examples for
which there is no point spectrum. For instance (see [183],
vol.II, p.986-989), if (φ_p) is a complete orthonormal system
in a compact interval $[a,b]$ of $\mathbb{R}$ and f a bounded measurable function in $[a,b]$, one defines a bounded quadratic
form $A(x,x) = \sum\limits_{p,q} a_{pq}x_p x_q$ by the formulas

(9)
$$a_{pq} = \int_a^b f(\mu)\varphi_p(\mu)\varphi_q(\mu)d\mu$$

and one has the "reduction" formulas

(10)
$$(x|x) = \int_a^b \left(\sum_p x_p\varphi_p(\mu)\right)^2 d\mu$$
$$A(x,x) = \int_a^b f(\mu)\left(\sum_p x_p\varphi_p(\mu)\right)^2 d\mu$$

from which it is easy to see that there is no point spectrum
(unless f is constant in an interval), and if f is continuous, the continuous spectrum consists of the whole interval
$[m,M]$, where m and M are the minimum and maximum value
taken by f in $[a,b]$. If one takes $a = 0$, $b = \pi$,
$\varphi_p(\mu) = \sqrt{\frac{2}{\pi}} \sin p\mu$, $f(\mu) = \cos \mu$, one obtains the first
example given by Hilbert

(11) $A(x,x) = x_1 x_2 + x_2 x_3 + x_3 x_4 + \cdots .$

If one takes $a = -\pi$, $b = \pi$, $\varphi_p(\mu) = \frac{1}{\sqrt{2\pi}}(\sin p\mu + \cos p\mu)$
for $-\infty < p < +\infty$, and $f(\mu) = -\pi - \mu$ if $\mu < 0$, $f(\mu) = \pi - \mu$

if $\mu > 0$, one gets $a_{pq} = \dfrac{1}{p+q}$ if $p+q \neq 0$, $a_{pq} = 0$ if

$q = -p$. Making $x_p = 0$ for $p \leq 0$, one gets the second

example of Hilbert $A(x,x) = \underset{p,q=1}{\Sigma} \dfrac{x_p x_q}{p+q}$, and making $x_p = 0$

for $p \leq 0$, $y_p = 0$ for $p \geq 0$, one gets still another

example of Hilbert $A(x,y) = \underset{p,q}{\Sigma}' \dfrac{x_p y_q}{p-q}$ (where the summation

extends to the pairs (p,q) of integers >0 and distinct);

both have for continuous spectrum the interval $[-\pi,+\pi]$ [*].

   In his 1913 book ([183],vol.II,p.956-989 and [184]), F.Riesz

gave an exposition of Hilbert's results based on an entirely

---

[*] One also should mention how Stieltjes' results on continued
fractions and on the moment problem were soon recognized as
belonging to the Hilbert-von Neumann spectral theory. Jacobi,
in 1848, had considered the special quadratic forms $J(x) =$
$= \overset{n}{\underset{k=1}{\Sigma}} a_k x_k^2 - 2 \overset{n-1}{\underset{k=1}{\Sigma}} b_{k+1} x_k x_{k+1}$, and he had shown that the eigenvalues of that form are the roots of the denominator of the
limited continued fraction ([120],vol.VI,p.318-321)

(F) $\qquad \dfrac{1}{a_o-z} - \dfrac{b_1^2 \,|}{|a_1-z} - \dfrac{b_2^2 \,|}{|a_2-z} - \cdots - \dfrac{b_n^2 \,|}{|a_n-z}$.

Already in 1878, Heine [104,vol.I. p.421] had hinted at the
possibility that an <u>unlimited</u> continued fraction of type (F)

would be similarly related to a Jacobi quadratic form

$\overset{\infty}{\underset{n=0}{\Sigma}} a_n x_n^2 - 2 \overset{\infty}{\underset{n=0}{\Sigma}} b_{n+1} x_n x_{n+1}$ in an infinite system of varia-

bles. What Stieltjes had done was to study directly unlimited
continued fractions of type (F), representing them as
"Stieltjes transforms" and being led to his "problem of
moments" by the problem of determining the Stieltjes measure
corresponding to a given Stieltjes transform; but he had not
considered the relation between the continued fraction and
quadratic forms. This was done by Toeplitz in 1910 [214] for
the case of Jacobi bounded quadratic forms, and later extended
to the general case; it turned out that these forms exactly
corresponded to spectra of <u>multiplicity</u> 1 [207].

different method, and which was to remain standard until
around 1950. As we have already pointed out, he replaces the
bilinear forms of Hilbert by the much more natural continuous
endomorphisms of $E = \ell_{\mathbb{R}}^2$; to such an endomorphism A is as-
sociated the "bounded" bilinear form $(x,y) \mapsto (A \cdot x | y)$ and
conversely each such form can be uniquely written in that way.
F. Riesz's central idea is to define, for such endomorphisms
A , "functions" $f(A)$ which would again be continuous endomorphisms of E, for suitable functions f of a real variable, and to use such "functions" to write for a symmetric en-
domorphism A (i.e. such that $(A \cdot x | y) = (x | A \cdot y)$ for all
x,y in E) a canonical "spectral decomposition" corresponding to formulas (8) of Hilbert.

   To develop these ideas, F. Riesz begins by some general results on the algebra $\mathcal{L}(E)$ of all continuous endomorphisms
of E. For the norm $\|A\| = \sup_{\|x\|=1} \|A \cdot x\|$, it is a Banach space,
has a unit (the identity mapping 1_E) and is such that
$\|AB\| \leq \|A\| \cdot \|B\|$. However, F. Riesz, for his purpose, is led
to use, not the notion of ("uniform") convergence derived
from the norm of $\mathcal{L}(E)$, but the notion of strong convergence:
he says a sequence (A_n) converges strongly to A if, for
every $x \in E$, the sequence $(\|A_n \cdot x - A \cdot x\|)$ tends to 0.

   F. Riesz then considers the subspace $\mathcal{H}(E)$ of all symmetric
operators; there is in $\mathcal{H}(E)$ an order relation, $A \leq B$ mean-
ing that $(A \cdot x | x) \leq (B \cdot x | x)$ for all $x \in E$. Suppose $a \cdot 1_E \leq$
$\leq A \leq b \cdot 1_E$; then, for every polynomial $P(\xi)$ with real coefficients such that $P(\xi) \geq 0$ in the interval $[a,b]$, one has
$P(A) \geq 0$. Indeed, one may assume $a = 0$, $b = 1$, and then P

is a sum of polynomials of one of the types $Q(\xi)^2$, $\xi(Q(\xi))^2$, $(1-\xi)(Q(\xi))^2$ or $\xi(1-\xi)(Q(\xi))^2$, and it is enough to prove that $A(1_E-A)$ is ≥ 0; but from the Cauchy-Schwarz inequality for the positive quadratic form $(A \cdot x|x)$ it follows that $\|A \cdot x\|^4 \leq (A \cdot x|x)(A^2 \cdot x|A \cdot x) \leq \|x\|^2 \|A \cdot x\|^2$; this first implies that $\|A\| \leq 1$, and then $\|A \cdot x\|^4 \leq (A \cdot x|x)\|A \cdot x\|^2$ and finally $(A^2 \cdot x|x) \leq (A \cdot x|x)$. From this it follows at once that if a polynomial $P(\xi)$ with real coefficients is such that $m \leq P(\xi) \leq M$ in $[a,b]$, then one has $m \cdot 1_E \leq P(A) \leq M \cdot 1_E$, and $\|P(A)\| \leq \sup(|m|,|M|)$.$^{(*)}$

These results first imply that if a sequence (P_n) of polynomials converges uniformly to a continuous function f in $[a,b]$, then the sequence $(P_n(A))$ is a Cauchy sequence in the Banach space $\mathcal{L}(E)$, and its limit only depends on f, hence can be written $f(A)$; furthermore the mapping $f \mapsto f(A)$ is a homomorphism of the algebra $C([a,b])$ into $\mathcal{L}(E)$, with values in $\mathcal{H}(E)$, which justifies the notation; in addition, if $m \leq f(\xi) \leq M$ in $[a,b]$, one has again $m \cdot 1_E \leq f(A) \leq$ $\leq M \cdot 1_E$.

But F. Riesz goes further. If (f_n) is an increasing sequence of continuous functions in $[a,b]$, uniformly bounded, then for any $x \in E$, the sequence of the $(f_n(A) \cdot x|x)$ is increasing and bounded, hence has a limit, from which it follows by linearity that the sequence of the $(f_n(A) \cdot x|y)$ converges for any pair of elements x, y in E; the Hellinger-

---

$^{(*)}$This argument is not the one used by Riesz, who deduces the result by a passage to the limit from the known result for the "Abschnitte" A_n of A.

Toeplitz theorem (chap.V, §4) shows that the limit can be written $(B \cdot x | y)$ where $B \in \mathcal{H}(E)$; if g is the (simple) limit of the sequence (f_n), one writes again $B = g(A)$, and this enables one to define $g(A)$ for any bounded (upper or lower) semi-continuous function in $[a,b]$ or any linear combination of such functions, which again form an algebra and for which $g \mapsto g(A)$ is a homomorphism.

F. Riesz then uses these results to obtain the spectral decomposition in the following way; if e_ξ is the function defined in $\mathbb{R}$ and such that $e_\xi(\mu) = 1$ for $\mu < \xi$, $e_\xi(\mu) = 0$ for $\mu \geq \xi$, $e_\xi(A) = A_\xi$ is defined since e_ξ is bounded and lower semi-continuous. For any pair of vectors x, y the function $\xi \mapsto (A_\xi \cdot x | y)$ is then a function of bounded variation, and for any continuous function f, one has

$$(12) \qquad (f(A) \cdot x | y) = \int_{-\infty}^{+\infty} f(\xi) \, d(A_\xi \cdot x | y)$$

a formula which one also writes

$$(13) \qquad f(A) = \int_{-\infty}^{+\infty} f(\xi) \, dA_\xi$$

and which is justified by the fact that for any $\epsilon > 0$, it is possible to divide the interval $[a,b]$ by points ξ_k in such a way that

$$-\epsilon \cdot 1_E \leq f(A) - \sum_k f(\mu_k)(A_{\xi_{k+1}} - A_{\xi_k}) \leq \epsilon \cdot 1_E$$

$$(\text{with } \xi_k \leq \mu_k \leq \xi_{k+1})$$

The spectrum[*] of A, contained in the interval $[a,b]$, is

_____

[*]To pass from the Hilbert notion of "spectrum" to the one used by F. Riesz, one must replace the parameter λ of Hilbert by $1/\xi$.

the complement of the set of points having a neighborhood
where $\xi \mapsto A_\xi$ is constant.

The operator A_ξ is the <u>orthogonal projector</u> of E onto a
closed subspace E_ξ, which is stable under A and in which
$(A \cdot x | x) < \xi (x | x)$ for all $x \neq 0$; $1_E - A_\xi$ is the orthogonal
projector of E onto the subspace $E_\xi^\perp$ orthogonal to E_ξ,
and in which $(A_\xi \cdot x | x) \geq \xi (x | x)$. The <u>point spectrum</u> of A is
the denumerable set of values of ξ where at least one of
the functions $\xi \mapsto (A_\xi \cdot x | y)$ is discontinuous; it consists of
all the <u>eigenvalues</u> of A, but the subspace N_ξ formed by
the corresponding eigenvectors may have infinite dimension.
The subspaces E_ξ are such that $E_\xi \subset E_\eta$ for $\xi < \eta$,
$E_\xi = \{0\}$ for $\xi < a$, $E_\xi = E$ for $\xi > b$; the intersection
of the subspaces E_η for $\xi < \eta$ is reduced to E_ξ if ξ is
not in the point spectrum and is the direct sum of the (orthogonal) subspaces E_ξ and N_ξ if ξ is in the point spectrum.

If $(A \cdot x | y) = \sum_{i,k} a_{ik} x_i y_k$ is the "bounded" bilinear form
corresponding to the operator A, the eigenvectors $x = (x_k)$
corresponding to an eigenvalue λ have coordinates which are
solutions of the system of linear equations

$$(14) \qquad \lambda x_i - \sum_k a_{ik} x_k = 0 \qquad (i=1,2,\ldots) .$$

However, Hilbert's theory left unanswered the question of determining "objects" which would replace the eigenvectors when
a number λ in the spectrum was not in the point spectrum.
In some cases this question had a curious answer; for instance, for the form (11), where there is no point spectrum and
the spectrum is the interval $[-1,1]$, for each value $\lambda = \cos t$

in that interval, the system (14) does have a solution, namely $x_k(t) = \sin kt$ for $k = 1, 2, \ldots,$ as it is easily ve-
rified; but for such a sequence, the series $\sum\limits_{k} x_k(t)^2$ is <u>not</u>
<u>convergent</u>. The existence of such "generalized eigenvectors"
was only incorporated in a general theorem much later (see
chap. IX, §2); but in the case of the form (11), one could
observe that if one wrote $\rho_k(\xi) = \int_0^\xi x_k(t)dt,$ then the
vector $(\rho_k(\xi))_{k=1, 2, \ldots}$ this time belonged to E for all
$\xi \in \mathbb{R},$ and it followed from the relations (14) satisfied by
the $x_k(t)$ that one could write, for <u>any interval</u> $[\lambda_1, \lambda_2]$
of $\mathbb{R}$

$$(15) \quad \int_{\lambda_1}^{\lambda_2} \xi\, d\rho_i(\xi) - \sum_k a_{ik}(\rho_k(\lambda_2) - \rho_k(\lambda_1)) = 0 \quad (i = 1, 2, \ldots).$$

This led Hellinger [105] to study systematically the sequences of functions $(\rho_k(\xi))_{k=1, 2, \ldots}$ of bounded variation which
satisfied (15) for all intervals and for which $\sum\limits_{k} \rho_k(\xi)^2$ is
finite for all ξ; he called the $d\rho_k$ "eigendifferentials"
of the operator A. This study allowed him to attack a problem which naturally arose from Hilbert's spectral theory, by
analogy with the finite dimensional case: can one give necessary and sufficient conditions for two operators A, B with
symmetric matrices to be "similar", i.e. such that there is
an orthogonal transformation U of E onto itself such that
$B = UAU^{-1}$? In the finite dimensional case, the condition is
that the eigenvalues of A and B be the same, with the same
multiplicity for each; the combined efforts of Hellinger and
H. Hahn [95] succeeded in obtaining necessary and sufficient
conditions for operators in $\ell^2_{\mathbb{R}}$, expressed in terms of

special systems of "eigendifferentials". We shall not give
here the detail of these complicated conditions, which we
shall formulate in a much simpler way using the Gelfand theory
of commutative Banach algebras (§5).

One of the byproducts of F. Riesz's method is that it enabled
him to give a <u>direct</u> definition of the whole spectrum of A ,
without any reference to the decomposition (12): a point
$\lambda \in \mathbb{R}$ is in the spectrum if and only if the operator $\lambda \cdot 1_E - A$
<u>does not have a continuous inverse</u>. Finally, he remarked that
his method could (just as Hilbert's method) be extended to
<u>self-adjoint</u> bounded operators A in <u>complex</u> Hilbert space $\ell_{\mathbb{C}}^2$,
i.e. those which satisfy the same condition $(A \cdot x | y) = (x | A \cdot y)$
for the hermitian scalar product in that space; the mapping
$\zeta \mapsto (\zeta \cdot 1_E - A)^{-1}$ is then <u>holomorphic</u> outside of the spectrum
of A . After 1913, almost all papers on spectral theory in
Hilbert space dealt exclusively with complex Hilbert space.

§3 - The work of Weyl and Carleman

Hilbert's method associating to an integral equation with
symmetric kernel $K(s,t)$ a "bounded" bilinear form $K(x,y)$
(chap.V, §2) worked even if $K(s,t)^2$ was not integrable in
$[a,b] \times [a,b]$, but the corresponding bilinear form might not
be "completely continuous"; already in his lectures in 1906
Hilbert had given the example $K(s,t) = (s+t)^{-1}$ for the interval $[0,1]$ ([227], vol.I, p.83). He also had observed in
his Seminar that the Fredholm theorems might fail when the
interval $[a,b]$ was unbounded, and had given as example an

interpretation of the Fourier inversion formula (1) (ibid.,
p.2): for $K(s,t) = \cos st$ in the interval $[0,+\infty[$, $\sqrt{\frac{2}{\pi}}$
and $-\sqrt{\frac{2}{\pi}}$ are the only eigenvalues, but each of them has
infinite multiplicity[(*)]. He therefore encouraged his most
gifted student, Hermann Weyl, to elucidate such "singular"
integral equations, and in particular to determine conditions
on the kernel $K(s,t)$ implying that the bilinear form $K(x,y)$
would be "bounded" and therefore amenable to his spectral
theory. This was the theme of Weyl's dissertation; in it and
in a subsequent paper (ibid., p.2-86 and 102-153) he gave
the following sufficient conditions for $[a,b[= [0,+\infty[$:

1º for each $s \geq 0$, the integral $\int_0^\infty (K(s,t))^2 dt$ is finite;

2º there is a constant $M > 0$ such that the inequality

$|\int_0^\infty \int_0^\infty K(s,t)u(s)v(t)dsdt| \leq M$ holds for all pairs of continuous

functions u, v such that $\int_0^\infty (u(s))^2 ds \leq 1$ and $\int_0^\infty (v(s))^2 ds \leq$
≤ 1.

Another direction of research derived from an interpretation of the Sturm-Liouville theory (chap.I, §3) in terms of
integral equations. Consider a second order differential
equation

(16) $y'' - q(x)y + \lambda y = f(x)$

where q and f are continuous functions in a compact interval $[a,b]$, q having real values, f real or complex
values and λ is a complex parameter; in addition we have

[(*)]This is apparently the first appearance of a relation
between Fourier transforms and Functional Analysis (see §6).

two boundary conditions

(17) $y(a)\cos \alpha - y'(a)\sin \alpha = 0, \quad y(b)\cos \beta - y'(b)\sin \beta = 0$

where α and β are two positive constants. An elementary
argument shows that for $f = 0$, the homogeneous equation (16)
with boundary conditions (17) has no solution if $\lambda \leq -r$,
where r is a number >0 depending only on q. Replacing
$q(x)$ by $q(x) + r$ and λ by $\lambda + r$, one may assume that for
$\lambda \leq 0$, the homogeneous equation $y'' - q(x)y + \lambda y = 0$ has no
solution satisfying the conditions (17).

Now consider first equation (16) for $\lambda = 0$; there are two
solutions u_1, u_2 of the equation $y'' - q(x)y = 0$ such that
$u_1(a)\cos \alpha - u_1'(a)\sin \alpha = 0, \quad u_2(b)\cos \beta - u_2'(b)\sin \beta = 0,$
u_1 and u_2 being <u>linearly independent</u>. For each $t \in [a,b]$,
define the function

(18) $\begin{cases} K(t,x) = -u_2(t)u_1(x)/d & \text{for} \quad a \leq x \leq t \\ K(t,x) = -u_1(t)u_2(x)/d & \text{for} \quad t \leq x \leq b \end{cases}$

where d is the constant $u_1(x)u_2'(x) - u_2(x)u_1'(x)$. The
function $x \mapsto K(t,x)$ is then continuous in $[a,b]$; in each
of the semi-open intervals $a \leq x < t$, $t < x \leq b$ (for
$a < t < b$) it satisfies the equation $y'' - q(x)y = 0$, and in
addition it satisfies conditions (17); finally, the function
$x \mapsto \frac{\partial}{\partial x} K(t,x)$ has at the point $x = t$ a discontinuity such
that $\frac{\partial}{\partial x} K(t,t+) - \frac{\partial}{\partial x} K(t,t-) = -1$. A routine calculation
then shows that in order that a function y be a solution of
$y'' - q(x)y = f(x)$ satisfying conditions (17), it is necessary
and sufficient that

$$(19) \qquad\qquad y(x) = - \int_a^b K(t,x)f(t)dt$$

and therefore the solutions of (16), satisfying (17), are
exactly the solutions of the integral equation

$$(20) \qquad\qquad y(x) - \lambda \int_a^b K(t,x)y(t)dt = g(x)$$

where

$$(21) \qquad\qquad g(x) = - \int_a^b K(t,x)f(t)dt.$$

Clearly this method was patterned after Schwarz's method for
solving the equation of vibrating membranes with the help of
the Green function for the Laplacian (chap.III, §1), and the
function K was therefore called the Green function for the
operator $L(y) = y'' - q(x)y$ and the boundary conditions (17)
[35]. As obviously $K(x,t) = K(t,x)$, the Sturm-Liouville
problem was thus reduced to a special case of the Fredholm-
Hilbert theory of integral equations with symmetric kernels.

In his second paper (1904) on integral equations, Hilbert
developed that method and expanded it to other boundary conditions than (17). He also was aware that many of the
"special functions" which had been introduced in Analysis
since the XVIIIth century (hypergeometric functions, Bessel
functions, Legendre polynomials, Hermite functions, etc.) satisfied equations of type (16) but with less restrictive conditions: the interval [a,b] would be replaced by an
unbounded interval and the function q might have singular
points at the extremities of the interval; in such a case,
Hilbert proposed that the corresponding boundary condition

should be replaced by the condition that y remain bounded
at such an extremity, or tends to infinity not faster than
some given singular function. He showed then, on various
examples, that one could even in such "singular" cases, define a "Green function" K, with the symmetric property
$K(t,x) = K(x,t)$, and the same discontinuity for the partial
derivatives for x = t; for instance, if $L(y) = y'' + y$, one
has, for the interval $]-\infty, +\infty[$, $K(t,x) = -\frac{1}{2} \sin|x-t|$. At
that time, he had not yet developed his theory of "bounded"
bilinear forms, so he limited himself to cases in which the
Green function K was a kernel to which the Fredholm theory
was applicable ([112], p.39-58). But of course he was aware
that in examples such as the one above, one would fall on
"singular" integral equations, and one of his students,
E. Hilb, wrote his "Habilitationschrift" in 1908 on the application of Hilbert's theory of "bounded" forms to two special cases of "singular" Sturm-Liouville problems [110].

Then, in 1909-1910, H. Weyl discovered that he could apply
the results of his dissertation to handle the most general
such problems for second order operators of the type

$$(22) \qquad L(y) = \frac{d}{dx} \left(p(x) \frac{dy}{dx} \right) - q(x)y$$

where p and q are real continuous functions in an interval
$I \subset \mathbb{R}$ (bounded or not) subject to the only restriction that
$p(x) > 0$ in I. In his "Habilitationschrift" ([227], vol.I,
p.248-297) he considers the case $I = [0, +\infty[$; his very original method consists in studying the equation

$$(23) \qquad\qquad L(y) + \lambda y = 0$$

for <u>non real</u> values of λ, and in fact, he sees that it is enough to consider the case $\lambda = i$. Let u_1, u_2 be the two solutions of $L(y) + iy = 0$ in $[0,+\infty[$ satisfying the initial conditions

$$u_1(0) = 1, \quad p(0)u_1'(0) = 0, \qquad u_2(0) = 0, \quad p(0)u_2'(0) = 1.$$

Let α be any number ≥ 0, and consider the two solutions of $L(y) + iy = 0$

$$v_\alpha = -\sin \alpha \cdot u_1 + \cos \alpha \cdot u_2, \quad w_\alpha = \cos \alpha \cdot u_1 + \sin \alpha \cdot u_2$$

so that v_α is a solution satisfying the boundary condition

$$(24) \qquad \cos \alpha \cdot y(0) + \sin \alpha \cdot p(0)y'(0) = 0.$$

Now take any number $a > 0$, a number $\beta \geq 0$, and determine the complex number μ by the condition that the solution $u_\mu = v_\alpha + \mu w_\alpha$ satisfies the boundary condition

$$(25) \qquad \cos \beta \cdot y(a) + \sin \beta \cdot p(a)y'(a) = 0.$$

Weyl shows that the uniquely determined number μ describes a <u>circle</u> Γ_a in the upper half plane $\mathcal{I}z > 0$ when β varies from 0 to 2π.

From the two solutions v_α and u_μ one can form a <u>Green function</u> $K_a^\mu(t,x)$ in the interval $[0,a]$ by the same formulas as (18), except that now K_a^μ has complex values. Now let a <u>tend to</u> $+\infty$; Weyl shows that the circles Γ_a form a nested family of decreasing radius, hence have a <u>limit</u> Γ_∞ which may be either a circle of radius >0, or a single point. In any case, if one lets the points $\mu \in \Gamma_a$ tend to a limit $\mu_0 \in \Gamma_\infty$, u_μ tends to a solution u_{μ_0} and K_a^μ tends to a

kernel $K^{\mu_0}(t,x)$ which always satisfies the two conditions
of Weyl's dissertation; this enables one to apply to the corresponding singular integral equation Hilbert's theory of
"bounded" bilinear forms. Just as for the usual Sturm-Liouville problem the function

$$(26) \qquad y(x) = - \int_0^\infty K^{\mu_0}(t,x)\hat{f}(t)dt$$

is then a solution of

$$(27) \qquad L(y) + iy = f(x)$$

satisfying the boundary condition (24) at the extremity 0,
and the condition "at infinity"

$$(28) \qquad \lim_{t\to+\infty} p(t)(y(t)u'_{\mu_0}(t) - u_{\mu_0}(t)y'(t)) = 0.$$

Furthermore, the solution u_{μ_0} always <u>belongs to</u> $L^2(0,\infty)$.
Weyl then shows that the problem splits in two cases:

I) The "<u>limit circle</u>" case, Γ_∞ being a circle of radius
>0. Then <u>all</u> solutions of $L(y) + iy = 0$ belong to $L^2(0,\infty)$,
and for all $\mu_0 \in \Gamma_\infty$, K^{μ_0} is a Hilbert-Schmidt kernel; the
conclusions of the Sturm-Liouville theory are then again valid
for equation (23) with the boundary condition (24) at extremity 0, and the condition that the real part of (28) vanishes
at extremity +∞.

II) The "<u>limit point</u>" case, when Γ_∞ is reduced to a single
point μ_0. Then u_{μ_0} is (up to a constant factor) the <u>only</u>
solution of $L(y) + iy = 0$ belonging to $L^2(0,\infty)$, and Weyl's
main objective is to set up integral formulas which should be
substituted to the "Fourier expansions" of the classical Sturm-

Liouville theory, as had been expected by Wirtinger, and
obtained by Hilb in special cases; this he is able to do by
applying the results of his dissertation to the "singular"
integral operator having as kernel the imaginary part of K^{μ_o},
and extensively using Hellinger's "eigendifferentials". He
also defines the spectrum of the differential operator L as
the complement of the set of parameters $\lambda \in \mathbb{R}$ for which the
differential equation $L(y) + \lambda y = g(x)$ has a solution belonging to $L^2(0,\infty)$ and satisfying (24), for every continuous
function g belonging to $L^2(0,\infty)$; he studies the structure
of that spectrum under various assumptions on the functions
p and q, and in particular he gives an example where that
spectrum is the whole real line. Finally, in a subsequent
paper ([227], vol.I, p.222-247), he shows how his theory may
be extended when the interval I is the whole line $\mathbb{R}$, and
the equation belongs to the "limit point" type at both extremities.

   Viewed from the vantage point of the later von Neumann
theory (see §4) these remarkable results of Weyl constitute
the first study of an unbounded hermitian operator in Hilbert
space, with non zero "defects", and of its self-adjoint extensions; the singular integral operator defined by the kernel
K^{μ_o} with complex values is probably the first example of a
normal operator in Hilbert space which is not self-adjoint[*].

---

[*]The concept of a normal square matrix A with complex elements had been defined in 1877 by Frobenius by the condition
that $AA^* = A^*A$; he had proved that this condition was equivalent to the existence of a unitary matrix U such that UAU^{-1}
is a diagonal matrix [78, vol.I, p. 391].

Such phenomena became even more apparent in the work of T. Carleman on singular integral equations ([37] and [38, p.313-342]) beginning in 1920. He starts from a kernel which only satisfies the _first_ of Weyl's assumptions, namely $\int_a^b |K(s,t)|^2 dt$ is almost everywhere finite (they are now called Carleman kernels, and the corresponding operators, which to f associate $s \mapsto \int_a^b K(s,t)f(t)dt$, Carleman operators). He treats the theory of these operators (for hermitian kernels) by a method similar to H. Weyl's: namely, he considers an increasing sequence (A_n) of measurable subsets of $[a,b]$, the union of which is equal to $[a,b]$ up to a null set, and which are such that he kernel $K_n(s,t)$, equal to $K(s,t)$ for $(s,t) \in A_n \times A_n$ and to 0 outside, has a finite integral $\int_a^b \int_a^b |K_n(s,t)|^2 dsdt$. If one now considers the integral equation

$$(29) \qquad \varphi(s) - \lambda \int_a^b K(s,t)\varphi(t)dt = f(s)$$

for _non real_ λ, f being in $L^2([a,b])$, one approximates it by the sequence of ordinary Hilbert-Schmidt integral equations

$$(30) \qquad \varphi_n(s) - \lambda \int_a^b K_n(s,t)\varphi_n(t)dt = f(s)$$

which (due to the choice of λ) have a unique solution $\varphi_n \in L^2([a,b])$. Carleman's original procedure is to integrate (30) after multiplication by $\overline{\varphi_n(s)}$, which gives him the identity between the imaginary parts

$$(31) \quad (\frac{1}{\lambda} - \frac{1}{\bar{\lambda}}) \int_a^b |\varphi_n(s)|^2 ds = \frac{1}{\lambda} \int_a^b \overline{\varphi_n(s)} f(s)ds - \frac{1}{\bar{\lambda}} \int_a^b \varphi_n(s)\overline{f(s)}ds$$

independent of the kernel, and from which he deduces, by the
Cauchy-Schwarz inequality,

$$(32) \qquad \int_a^b |\varphi_n(s)|^2 ds \leq \frac{4|\lambda|^2}{|\lambda - \bar{\lambda}|^2} \int_a^b |f(s)|^2 ds.$$

Applying the usual "principle of choice" and density arguments
as Hilbert and F. Riesz had done, Carleman is able (by letting
n tend to $+\infty$) to define a linear mapping $f \longmapsto T_\lambda \cdot f$ in
L^2 such that $\varphi = T_\lambda \cdot f$ is a solution of (29) for each $f \in L^2$,
and a passage to the limit in (32) shows that T_λ is contin-
uous. He even goes as far as writting an equation equivalent
to

$$(33) \qquad T_\lambda \cdot f = \frac{\bar{\lambda}}{\lambda - \lambda} \left(f + U_\lambda \cdot f \right)$$

and showing that $\| U_\lambda \cdot f \| \leq \| f \|$.

He then realized that the solution of (29) for non real λ
might be <u>non unique</u>, and he gave examples of kernels where
this phenomenon happens, which he called kernels of class II;
the other ones he called kernels of class I, and he showed
that they may be more general than the continuous operators
of F. Riesz or those considered by H. Weyl.

For any functions φ, ψ in L^2, the functions

$$(34) \quad S \cdot \varphi: s \mapsto \int_a^b K(s,t)\varphi(t)dt, \qquad S' \cdot \psi: s \mapsto \int_a^b K(t,s)\psi(t)dt$$

are always defined for a hermitian Carleman kernel K, but
the set D (resp. D') of functions of L^2 such that
$S \cdot \psi \in L^2$ (resp. $S' \cdot \psi \in L^2$) is in general a proper subspace
of L^2; D' consists of the complex conjugates of the func-
tions of D. Carleman showed that a necessary and sufficient

condition for K to be "of class I" was that the relation

$$(35) \qquad \int_a^b f(s)(S' \cdot \bar{g})(s)ds = \int_a^b \overline{g(s)}(S \cdot f)(s)ds$$

should hold for all f,g in D.

He next proceeded to let also n tend to +∞ in the Hilbert formula for Hilbert-Schmidt kernels (chap.V, §2, formula (24)) for the kernels K_n and their conjugates $\bar{K}_n$, and obtained for the operators T_λ and T'_λ corresponding to K and $\bar{K}$ formulas similar to those obtained by Hilbert in his theory of "bounded" quadratic forms. For kernels "of class II", the study of these formulas led Carleman to single out the case in which the operator U_λ in (33) and the corresponding operator U'_λ for T'_λ are both <u>unitary</u>; he shows that this property is independent of the choice of the non real number λ in one of the half planes $\Im\lambda > 0$, $\Im\lambda < 0$, and that it implies that the dimensions of the spaces of solutions of $\varphi - \lambda S \cdot \varphi = 0$ and $\psi - \lambda S' \cdot \psi = 0$ in L^2 are the <u>same</u>; finally he proved that in this case there are <u>infinitely</u> many operators T_λ and T'_λ having the above property (what he calls "maximal solutions") for the same kernel K.

All these results were quite surprising, in particular the existence of solutions $\varphi \neq 0$ for the equation $\varphi - \lambda S \cdot \varphi = 0$ in L^2 for non real λ, which seemed to contradict the classical argument (going back to Poisson (chap.I, §3), and even to Lagrange ([135], vol.XII, p.239) in the finite dimensional case) which, from the <u>reality</u> of the number $(S \cdot \varphi | \varphi)$ for all φ, concluded to the impossibility of a non real number λ satisfying $\varphi = \lambda S \cdot \varphi$ for $\varphi \neq 0$, since it implied

$(\varphi|\varphi) = \lambda(S \cdot \varphi|\varphi)$.   We shall see in the next section how this

apparent contradiction was resolved in the von Neumann theory,

which put the pioneering results of Carleman in their proper

context.

§4 - The spectral theory of von Neumann

In the fall of 1926, the young J. von Neumann (1903-1957)

arrived at Göttingen, to take up his duties as Hilbert's

assistant. These were the hectic years during which quantum

mechanics was developing at breakneck speed, with a new idea

popping up every few weeks from all over the horizon [121].

The theoretical physicists who were developing the new theory

were groping for adequate mathematical tools, trying in suc-

cession infinite matrices without any consideration of con-

vergence[*], differential operators, "continuous" matrices

(whatever that might mean) etc. It finally dawned upon them

that their "observables" had properties which made them look

like hermitian operators in Hilbert space, and that, by an

extraordinary coincidence, the "spectrum" of Hilbert (a name

which he had apparently chosen from a superficial analogy) was

to be the central conception in the explanation of the "spectra"

of atoms. It was therefore natural that they should enlist

Hilbert's help in trying to put some mathematical sense in

their computations. With the assistance of Nordheim and von

Neumann, he first tried integral operators in L^2, but that

needed the use of the Dirac "δ-function", a concept which for

---

[*]As late as 1924, most physicists did not even know what a

<u>finite</u> matrix was!

the mathematics of that time was self-contradictory (cf. chap.
VIII, §3); von Neumann therefore resolved to try another
approach.

Ever since the discovery of the Fischer-Riesz theorem (chap.
V, §3) the isomorphism of the space of sequences $\ell_{\mathbb{C}}^2$ and of
the $L_{\mathbb{C}}^2(\Omega)$ spaces of classes of quadratically integrable
functions in some subset Ω of an $\mathbb{R}^n$ had been familiar to
analysts, but by "Hilbert space" one always understood one of
these "concrete" spaces, on which the "operators" would therefore be, either "matrices" or "integral operators" of some
kind. Von Neumann was the first to conceive of an "abstract"
Hilbert space, defined axiomatically as a complex vector
space with a hermitian scalar product, separable and complete
for the corresponding norm, so that the usual "concrete"
Hilbert spaces would only be "incarnations" so to speak of
that "abstract" space. Obvious as it now seems to us, this
was a momentous step and opened the way to a complete unders-
tanding of spectral theory of normal and hermitian operators
in Hilbert space, which von Neumann proceeded to develop in
3 fundamental papers published between 1929 and 1932, and
which (with the exception of the description of the spectrum,
see §5) are still today, in substance, the definitive account
of the subject ([221], vol.II, p.1-85, 86-143 and 242-258)[*].

---

[*]During the same period, M.H. Stone, independently of von
Neumann, obtained the same results concerning self-adjoint
(unbounded) operators [206], and later gave a didactic exposition of the whole theory and of its applications at that time,
much clearer than von Neumann's papers, and which remained the
reference book on the subject for many years [207].

Abandoning any "concrete" presentation of Hilbert space, von Neumann was compelled to work <u>intrinsically</u>, using only notions which could directly be defined from the concepts enumerated in the axioms, to the exclusion of anything else. This led him to discover a remarkable series of entirely new ideas and methods.

1) Most operators used in quantum mechanics could <u>not</u> be defined in the whole Hilbert space, as for instance in $L^2(\mathbb{R})$ multiplication of functions by a fixed function such as $x \mapsto x$, or derivation of functions. One therefore had to consider, in general, linear mappings T taking their values in a Hilbert space E, but only defined in a <u>proper</u> vector subspace $\text{dom}(T)$ (the "domain" of T); the most interesting case concerned the operators T <u>densely defined</u>, i.e. those for which $\text{dom}(T)$ is dense in E (as in the two examples above).

2) If $\text{dom}(T)$ is dense in E and T is continuous, it can immediately be extended to the whole space E, and one is brought back to the Hilbert theory. But von Neumann had the idea to introduce a weaker substitute for continuity, namely the fact that the <u>graph</u> $\Gamma(T)$ of T be closed in $E \times E$; one then says that T is a <u>closed</u> operator. It is obvious that for $\text{dom}(T) = E$, if T is continuous, then T is closed, and the converse follows from the closed graph theorem (chap.VI, §5). Of the two examples given in 1), the first is closed but the second is not.

3) This last example raises the problem of <u>extending</u> (if possible) a densely defined operator T which is not closed to a closed operator; von Neumann was able to give a beautiful

anwer to that problem by linking it to a generalization of the
notion of <u>adjoint</u> operator, well known for bounded operators.
In general, for a densely defined operator T and a vector
$y \in E$, the linear form $x \mapsto (T \cdot x | y)$, defined in dom (T),
is not necessarily continuous; if it is, it can be extended
uniquely by continuity to E, and then can be written
$x \mapsto (x | y^*)$ for a unique vector $y^* \in E$; the set of vectors
$y \in E$ having that property is a vector subspace, and if one
writes $y^* = T^* \cdot y$ for those vectors, T^* is a linear operator defined in that subspace (which is therefore dom(T^*)).
Now it is easy to show that T^* is always closed (even if T
is not), and its graph is the subspace of E which is the
orthogonal supplement to the closure of $J(\Gamma(T))$, where J
is the linear automorphism $(x,y) \mapsto (y,-x)$ of ExE ("rotation
of a right angle"!). This interpretation of dom(T^*) gave
to von Neumann the proof of the <u>equivalence</u> of the two following properties: a) T can be extended to a closed operator (one
says T is <u>closable</u>); b) T^* is densely defined. One can
easily give examples in which dom(T^*) = {0}; if T is closable, the smallest closed extension of T is T^{**}, and one
has $\Gamma(T^{**}) = \overline{\Gamma(T)}$ and $(T^{**})^* = T^*$.

4) The fact that closed densely defined operators are not
everywhere defined raises difficulties concerning algebraic
combinations of such operators: $A + B$ is only defined in
dom(A) $\cap$ dom(B), AB only in dom(B) $\cap B^{-1}(\text{dom}(A))$; one
can give examples of closed densely defined operators T such
that dom(T^2) = {0}. However, using the decomposition of
ExE in the direct sum of the closed orthogonal subspaces

$\Gamma(T)$ and $J(\Gamma(T^*))$, von Neumann could prove that for any closed densely defined operator T, $\text{dom}(T^*T)$ is dense, T^*T is closed and $(T^*T)^* = T^*T$. Furthermore, $1_E + T^*T$ (closed and defined in $\text{dom}(T^*T)$) is a bijection of $\text{dom}(T^*T)$ onto the whole space E, the inverse $B = (1_E + T^*T)^{-1}$ is a <u>bounded</u> self-adjoint and injective operator, the spectrum of which is contained in the interval $[0,1]$.

5) These results enabled von Neumann to completely elucidate the spectral theory of <u>normal</u> operators in E. By definition, they are the closed densely defined operators N such that $\text{dom}(N^*N) = \text{dom}(NN^*)$ and $N^*N = NN^*$. The most important normal operators are the <u>self-adjoint</u> operators (which von Neumann called "hypermaximal"), defined by the condition $N^* = N$ (implying of course $\text{dom}(N^*) = \text{dom}(N)$), and the <u>unitary</u> operators, which are bounded and such that $N^*N = 1_E$ (hence invertible and such that $N^{-1} = N^*$).

Now F. Riesz's definition of the spectrum of a bounded operator can be generalized for any <u>closed</u> operator T in E. One says a complex number ζ is a <u>regular value</u> for T if the operator $T - \zeta 1_E$ is a <u>bijection</u> of the subspace $\text{dom}(T)$ onto the whole space E and if the inverse mapping $R_T(\zeta)$ (also called the <u>resolvent</u> of T) is a <u>bounded</u> operator mapping E onto $\text{dom}(T)$; it is enough for that to know that $T - \zeta 1_E$ is injective, that its image L is dense in E, and the inverse mapping $(T - \zeta 1_E)^{-1}$ of L onto $\text{dom}(T)$ is continuous. The complement $\text{Sp}(T)$ of the set of regular values of T in $\mathbb{C}$ is by definition the <u>spectrum</u> of T, and the mapping

$\zeta \mapsto R_T(\zeta)$ of $\mathbb{C} - \mathrm{Sp}(T)$ into $\mathcal{L}(E)$ is <u>holomorphic</u>. For a number $\zeta \in \mathrm{Sp}(T)$, there are 3 possibilities:

1º $T - \zeta 1_E$ is not injective, which means there exists an $x \in \mathrm{dom}(T)$ such that $x \neq 0$ and $T \cdot x = \zeta x$, in other words ζ is an <u>eigenvalue</u> of T; one then says ζ belongs to the <u>point spectrum</u> of T.

2º $T - \zeta 1_E$ is injective and its image L is dense in E, but the inverse mapping $(T - \zeta 1_E)^{-1}$ is not continuous in L; then ζ is said to belong to the <u>continuous spectrum</u> of T.

3º $T - \zeta 1_E$ is injective, but its image L is not dense in E; one says ζ belongs to the <u>residual spectrum</u> of T.

For <u>normal</u> operators, there is <u>no residual spectrum</u>; self-adjoint operators are characterized as normal operators for which the spectrum is <u>contained in</u> $\mathbb{R}$, and unitary operators are normal operators for which the spectrum is <u>contained in the unit circle</u> $\mathbb{U}$: $|\zeta| = 1$.

6) Generalizing F. Riesz's presentation of the Hilbert spectral theory (§2), von Neumann shows that to every <u>self-adjoint</u> operator A in E is naturally associated a unique <u>decomposition of unity</u>. He means by that a family $\lambda \mapsto E(\lambda)$ of <u>orthogonal projectors</u> in E, depending on a real parameter λ, and such that:

1º $E(\lambda) E(\mu) = E(\mu) E(\lambda) = E(\lambda)$ for $\lambda \leq \mu$;

2º when $\lambda > \lambda_o$ tends to λ_o, $E(\lambda)$ tends to $E(\lambda_o)$ <u>strongly</u>; when λ tends to $-\infty$, $E(\lambda)$ tends strongly to 0, and when λ tends to $+\infty$, $E(\lambda)$ tends strongly to 1_E;

3º for any $x \in E$, the mapping $\lambda \mapsto \|E(\lambda) \cdot x\|^2$ increases from 0 to $\|x\|^2$ in $\mathbb{R}$; dom(A) is exactly the set of

$x \in E$ such that the Stieltjes integral

$$\int_{-\infty}^{+\infty} \lambda^2 \, d(\|E(\lambda) \cdot x\|^2)$$

is __finite__;

4º for any $x \in$ dom(A) and any $y \in E$, the function
$\lambda \longmapsto (E(\lambda) \cdot x \,|\, y)$ is a function of bounded variation, and one
has the expressions

(36) $(A \cdot x \,|\, y) = \displaystyle\int_{-\infty}^{+\infty} \lambda \, d((E(\lambda) \cdot x \,|\, y)),$ $(x \,|\, y) = \displaystyle\int_{-\infty}^{+\infty} d((E(\lambda) \cdot x \,|\, y))$

as Stieltjes integrals.

Conversely, for any family $\lambda \longmapsto E(\lambda)$ of orthogonal projectors satisfying 1º and 2º, conditions 3º and 4º __define__ a
self-adjoint operator A and its domain, to which the given
family is its decomposition of unity. The operator A is
bounded if and only if there is a compact interval $[\alpha, \beta]$
such that $E(\lambda) = 0$ for $\lambda < \alpha$ and $E(\lambda) = 1_E$ for $\lambda > \beta$.
The spectrum of A is the complement of the set of points
$\mu \in \mathbb{R}$ such that $E(\lambda)$ is constant in a neighborhood of μ,
and the point spectrum is the set of points μ such that
$E(\mu-)$ is distinct from $E(\mu)$.

For __unitary__ operators U, there is a similar result: to U
corresponds a unique decomposition of unity $\lambda \longmapsto E(\lambda)$ satisfying conditions 1º and 2º above, with $E(\lambda) = 0$ for
$\lambda < 0$ and $E(\lambda) = 1_E$ for $\lambda > 1$; condition 3º is then automatically satisfied, and the first relation (36) is replaced
by

(37) $(U \cdot x \,|\, y) = \displaystyle\int_0^1 e^{2i\pi\lambda} \, d((E(\lambda) \cdot x \,|\, y)).$

There is a similar "decomposition" for all <u>normal</u> operators, but we shall give it in a much simpler equivalent form in §5.

7) The most original part of von Neumann's work on spectral theory is his discovery and study of <u>hermitian</u> operators in Hilbert space E, as <u>distinct</u> from self-adjoint operators. A hermitian operator H is a densely defined operator such that $\operatorname{dom}(H) \subset \operatorname{dom}(H^*)$ and that the restriction of H^* to $\operatorname{dom}(H)$ is equal to H , in other words

$$(38) \qquad (H \cdot x \mid y) = (x \mid H \cdot y) \quad \text{for} \quad x,y \quad \text{in} \quad \operatorname{dom}(H),$$

and in particular $(H \cdot x \mid x)$ is a <u>real</u> number for all $x \in \operatorname{dom}(H)$. This implies that H is closable, and its closure H^{**} is again a hermitian operator; one may therefore restrict the study to <u>closed</u> hermitian operators. The new idea of von Neumann is to adapt to Hilbert space a device introduced in 1855 by Cayley to parametrize the orthogonal group: he had shown that, for an $n \times n$ skew-real symmetric matrix S , such that $\det(I+S) \neq 0$, $U = (I-S)(I+S)^{-1}$ was an orthogonal matrix, and any orthogonal matrix U such that $\det(I+U) \neq 0$ could be written in that way. Similarly, for a closed hermitian operator H , one has $\|x\|^2 \leq \|H \cdot x + ix\|^2$ for $x \in \operatorname{dom}(H)$, which implies that the closed operator $H + iI$ is injective in $\operatorname{dom}(H)$, and maps $\operatorname{dom}(H)$ on a <u>closed</u> subspace F in such a way that $(H+iI)^{-1}$ is continuous in F, and $V: y \longmapsto (H-iI)(H+iI)^{-1} \cdot y$ is an <u>isometry</u> of F on a closed subspace $V(F)$ of E. Conversely, if U is an isometry of a closed subspace F of E onto another closed subspace $U(F)$ such that the image G of F by $I-U$ is <u>dense</u> in E, then $I-U$ is a <u>bijection</u> of F onto G, and if, for $y \in G$, one writes $H \cdot y = i(I + U)(I - U)^{-1} \cdot y$, H is a closed hermitian operator such that $\mathrm{dom}(H) = G$ and $U = (H - iI)(H + iI)^{-1} = V$ defined above.

Furthermore, if E_H^+ is the orthogonal supplement of F in E, it is exactly the subspace of $\mathrm{dom}(H^*)$ consisting of the solutions of $H^* \cdot x = ix$; similarly, the orthogonal supplement E_H^- of $V(F)$ in E is the subspace of the solutions of $H^* \cdot x = -ix$ in $\mathrm{dom}(H^*)$, and $\mathrm{dom}(H^*)$ is the <u>direct sum</u> of the three subspaces $\mathrm{dom}(H)$, E_H^+ and E_H^-.

8) This method enables von Neumann to give a description of <u>all</u> hermitian operators H_1 which <u>extend</u> a given hermitian operator H . It is enough to describe the isometry V_1 which is the "Cayley transform" of H_1: one takes a closed subspace M of E_H^+ and an isometry W of M onto a closed subspace N of E_H^-; V_1 is then defined in the Hilbert sum $F_1 = E \oplus M$, equal to V in F and to W in M; $E_{H_1}^+$ is the orthogonal supplement of M in E_H^+ and $E_{H_1}^-$ the orthogonal supplement of N in E_H^-. Such a construction is of course only possible if $\mathrm{dim}(M) \le \mathrm{dim}(E_H^-)$. The dimensions d^+ of E_H^+ and d^- of E_H^- are called the <u>defects</u> of H ; examples may be given in which they take any integral value or are infinite.

Self-adjoint operators are by definition hermitian operators for which $H^* = H$, or equivalently those for which the defects are $(0,0)$. It follows at once from the preceding remarks that the closed hermitian operators which can be extended to self-adjoint operators are exactly those for which both defects are <u>equal</u> (finite or infinite); unless they are

both 0, there are infinitely many such self-adjoint extensions.

   To give an example of a closed hermitian operator of defects
$(1,0)$, von Neumann takes an orthonormal basis $(e_n)_{n\geq 0}$ of E,
and in E considers the closed hyperplane F orthogonal to
e_o, hence spanned by the e_n with $n \geq 1$; he denotes by U
the isometry of F onto E defined by $U \cdot e_n = e_{n-1}$ for
$n \geq 1$; it is easy to show that the image of F by $I - U$ is
dense in E, and therefore U is the Cayley transform of a
closed hermitian operator H having the required property.
Another (non closed) hermitian operator is given by taking
$E = L^2(I)$, where I is any interval in $\mathbb{R}$, and $H = i\frac{d}{dx}$,
which is defined in the subspace of E consisting of C^1
functions vanishing at both extremities of I and whose derivative is square integrable (or any subspace of that subspace which is still dense in $L^2(I)$, for instance the space
of C^∞ functions with compact support in I); it may then be
shown that the defects of H^{**} are $(1,1)$ if I is bounded,
$(1,0)$ if I is only bounded from above, $(0,1)$ if I is
bounded from below and $(0,0)$ if $I = \mathbb{R}$.

   As we already mentioned (§3) the results of H. Weyl on linear second order equations with real coefficients can easily
be interpreted in the von Neumann theory: the differential
operator L is hermitian; the defects of L^{**} are $(2,2)$ in
the "limit circle" case, and $(1,1)$ in the "limit point" case.
They prefigurated the general spectral theory of formally
self-adjoint linear differential equations which developed
around 1950 (see chap.IX, §3).

Similarly, Carleman's results are interpreted in the following way: for a Carleman kernel K, if one writes $k(s)^2 =$
$= \int_a^b |K(s,t)|^2 dt$, in order to get a hermitian operator, one
should <u>restrict</u> the operator S defined in §3 (formula (34))
to the subspace (dense in L^2) of functions f such that the
integral $\int_a^b k(s)|f(s)|ds$ is finite; S is then the <u>adjoint</u>
of that operator, which explains the existence of non trivial
solutions of $S \cdot \varphi = i\varphi$ in $D = \text{dom}(S)$, and shows that the
operator U_λ, suitably restricted, coincides with the
"Cayley transform" which von Neumann later defined in a more
general context.

One should finally mention that von Neumann took pains, in
a special paper $([221], \text{vol.II}, \text{p.144-172})$, to investigate
how hermitian operators might be represented by infinite matrices (to which many mathematicians, and even more physicists,
were sentimentally attached); he pointed out that if one
wanted to associate to a hermitian operator H a matrix (a_{mn})
by the usual rule $a_{mn} = (H \cdot e_m | e_n)$ for an orthonormal basis
(e_n) of Hilbert space, one immediately ran into difficulties,
since the vectors $H \cdot e_m$ should be defined, in other words
one should have $e_n \in \text{dom}(H)$ for all n; furthermore, the
sums $\sum_n |a_{mn}|^2$ should all be finite. But if H is not maxi-
mal (i.e. both defects are >0), any hermitian operator which
extends H obviously has the <u>same</u> matrix (a_{mn}); and von
Neumann showed in great detail how this lack of "one-to-oneness"
in the correspondence between matrices and operators led to
the weirdest pathology, convincing once for all the analysts
that matrices were a totally inadequate tool in spectral theory.

§5 - Banach algebras

We have seen (§2) that F. Riesz probably was the first mathematician to consider the algebra $\mathcal{L}(E)$ of all continuous
endomorphisms of a separable Hilbert space E, with its norm
and what later came to be called its strong topology. In his
second paper on spectral theory ([221], vol.II, p.86-143) in
which he introduced the concept of normal operator in its most
general form, von Neumann began a more detailed study of $\mathcal{L}(E)$
and its subalgebras. He introduced the weak topology on $\mathcal{L}(E)$
(see chap. VIII, §1), and (inspired by the work of I. Schur
on linear representations of groups) the concept of commutant
M' of a subset M of $\mathcal{L}(E)$, but with an additional condition: M' should consist of all operators A such that, not
only A but also A^*, was permutable with all elements of M.
He focused his interest on the subalgebras of $\mathcal{L}(E)$ (later
called involutive or *-subalgebras) which, with every element
A, also contained its adjoint A^*; and he proved in that
paper the first two non trivial results on such subalgebras:
the double commutant M" of any involutive subalgebra M of
$\mathcal{L}(E)$ containing 1_E is the weak closure of M, and any
weakly closed commutative subalgebra of $\mathcal{L}(E)$ is generated
by a single self-adjoint operator A. A little later he completed this last result by showing that one could define
"functions $f(A)$" of a self-adjoint operator A for all universally measurable bounded functions f defined in $\mathbb{R}$ (and
not only for semi-continuous functions, as F. Riesz had done),
and he proved that the weakly closed subalgebra generated by

A consisted of all operators $f(A)$ thus defined ([221],

vol.II, p.177-212).

But for von Neumann this was only a beginning. The period

1926-1932 had seen the blossoming forth of the theory of

"hypercomplex numbers" of Molien, E. Cartan and Wedderburn

into the beautiful theory of "rings with descending chain con-

ditions" of E. Artin and E. Noether, followed by their appli-

cations to linear representations of groups and number theory

by R. Brauer, H. Hasse and A. Albert. Von Neumann was very

much interested by these developments, and wondered if it

could not be possible to build up some similar theory for in-

volutive subalgebras of $\mathcal{L}(E)$,   where of course "chain condi-

tions" could not be expected, but suitable topological res-

trictions would be a substitute, allowing one to obtain a

reasonable classification (loc.cit.,p. 89). It would take us

too far away from our main theme to describe in some detail

the series of papers, beginning in 1935, in which, with the

partial collaboration of F. Murray, he achieved a great part

of this program for what we now call the von Neumann algebras,

namely the involutive subalgebras equal to their double com-

mutant in $\mathcal{L}(E)$. By the wealth and novelty of their techniques

and their results, these wonderful papers are certainly the

most profound and most difficult which von Neumann ever wrote

([221], vol.III); they revealed a large number of completely

unsuspected phenomena, the most conspicuous one being the ap-

pearance, in the classification of the von Neumann algebras

with trivial center (those called factors), of five types of

algebras labeled I_n, I_∞, II_1, II_∞ and III,   where type I_n

means algebras of n×n matrices, type I_∞ the algebra $\mathcal{L}(E)$
itself, but the three other types were entirely unexpected
and exhibit new features, such as the attribution to the projectors contained in these algebras of a "dimension" which,
for algebras of Type II, may be <u>any real number</u> (in [0,1] or
[0,+∞]) instead of an integer. The elucidation of the properties of these new algebras, begun by Murray and von Neumann,
has engaged many mathematicians during the last 40 years, and
it is only recently that some difficult questions, such as the
classification of algebras of type III, have begun to be understood (see [57], [210] and [44]). Furthermore, since 1950
the von Neumann algebras have been an important tool in the
theory of linear representations of locally compact groups
(see §6); more recently they have been associated to foliations
and to generalizations of the Atiyah-Singer index (see [58],
[10], [31], [36], [45], [113], [132], [199]).

Surprisingly enough, the difficult theory of von Neumann
algebras was developed 5 years before the elementary concepts
of the theory of normed algebras had been defined! The cre-
ation of that theory was the work of I. Gelfand in 1941 [83];
a normed algebra A (over the <u>complex</u> field) is an algebra
over $\mathbb{C}$ on which is defined a structure of normed space with
the condition that the mapping $(x,y) \longmapsto xy$ of A×A into A
be continuous. It is then possible to choose on A a norm
compatible with the vector space structure and the topology
of A, and such that in addition $\|xy\| \leq \|x\| \cdot \|y\|$. If A has
a unit element e, one may suppose in addition that $\|e\| = 1$.
If A has no unit element, it is always possible to imbed A

into a normed algebra $\tilde{A}$ with a unit element e, such that
$\tilde{A}$ is the direct sum of A and $\mathbb{C}e$. For any normed space E
over $\mathbb{C}$, the algebra $\mathcal{L}(E)$ of endomorphisms of E is a
normed algebra for the norm $\|A\| = \sup\limits_{\|x\| \leq 1} \|A \cdot x\|$, but there are
many other types of normed algebras, the most elementary one
being the algebra $\mathcal{C}(I)$ of complex continuous functions in a
compact interval I of $\mathbb{R}$, with $\|f\| = \sup\limits_{t \in I} \|f(t)\|$.

Gelfand's main idea, which proved extraordinarily fruitful,
was to extend spectral theory to elements of normed algebras;
if A is a normed algebra with unit element e, one may
apply F.Riesz's definition of the spectrum (§2) to define the
spectrum of an arbitrary element $x \in A$: it is the set of
complex numbers ζ such that $x - \zeta e$ is not invertible in A.
Gelfand recognized that to get substantial results one must
assume that A is complete as a Banach space, what is called
a Banach algebra. Then very elementary arguments show that
the spectrum $\mathrm{Sp}_A(x)$ of any element $x \in A$ is a non empty
compact subset, contained in the disc $|\zeta| \leq \|x\|$; the inver-
tible elements in A form an open group G containing the
ball $\|x-e\| < 1$, and the topology induced on G is compatible
with the group structure; for any $x \in A$, the map $\zeta \mapsto (x-\zeta e)^{-1}$
of the complement $\mathbb{C}-\mathrm{Sp}_A(x)$ into A is holomorphic. Finally, Gelfand obtained a beautiful formula for the radius of the
smallest disc of center O containing $\mathrm{Sp}_A(x)$; this number,
called the spectral radius of x, is equal to

$$(39) \qquad \rho(x) = \lim_{n \to \infty} (\|x^n\|^{1/n}).$$

Next Gelfand undertook the study of general commutative

Banach algebras by a very original method. Probably inspired
by the theory of commutative groups (see §6), he defined a
character χ of a Banach algebra A as a homomorphism of
that algebra in the field $\mathbb{C}$ (considered as $\mathbb{C}$-algebra) which
is not identically O. Suppose for simplicity that A has a
unit element e; then any character χ is such that $\chi(e) = 1$,
and is a continuous linear form on A, of norm $\|\chi\| = 1$;
furthermore, for each $x \in A$, one has $\chi(x) \in \mathrm{Sp}_A(x)$.
Gelfand then associates to A the set $X(A)$ of all charac-
ters of A; the map $\chi \longmapsto \chi^{-1}(0)$ is a bijection of $X(A)$ on
the set of all maximal ideals in A (which are automatically
closed). Now, in 1937, Stone [208] had already considered
the set of maximal ideals of a very special type of ring, a
"boolean ring" B, which is commutative and such that $x^2 = x$
and $2x = 0$ for all $x \in B$; this kind of ring itself had
been suggested to Stone by the set of characteristic functions
φ_M of subsets M of an arbitrary set E, where multiplication is the usual one, and addition $\overset{\cdot}{+}$ is defined by
$\varphi_M \overset{\cdot}{+} \varphi_N = \varphi_{M \cup N} - \varphi_{M \cap N}$; furthermore, Stone of course was well
aware that for a self-adjoint operator A in Hilbert space,
the orthogonal projectors $\varphi_M(A)$ for universally measurable
subsets M of $\mathbb{R}$ form a boolean ring for the same addition.
The consideration of the set of maximal ideals of a commutative Banach algebra was therefore not at all foreign to the
spirit of spectral theory at that time.

As the set $X(A)$ is contained in the unit ball $\|x'\| \leq 1$
of the dual A' of the Banach space A, the natural embedding
of A into its second dual A'' associates to each element

$x \in A$ the map $\chi \mapsto \chi(x)$ of $X(A)$ into $\mathbb{C}$, which is called the Gelfand transform of x and is written $\mathcal{G}x$. It is easy to see that $X(A)$ is compact for the weak topology of A', and that $\mathcal{G}x$ is a continuous function on $X(A)$ for that topology; one has therefore defined a continuous homomorphism $x \mapsto \mathcal{G}x$ of the Banach algebra A into the Banach algebra $\mathcal{C}(X(A))$, such that the set of values of $\mathcal{G}x$ is the spectrum of x, and therefore $\|\mathcal{G}x\| = \rho(x) \leq \|x\|$. The compact space $X(A)$ is therefore called the spectrum of the Banach algebra A.

If one starts from the Banach algebra $A = \mathcal{C}(K)$ of continuous functions on a compact space K, then it is easy to see that $x \mapsto \mathcal{G}x$ is an isomorphism of A onto $\mathcal{C}(X(A))$, $X(A)$ being identified with the space of Dirac measures on K. But in general the homomorphism $x \mapsto \mathcal{G}x$ of A into $\mathcal{C}(X(A))$ is neither surjective nor injective. A little later, in collaboration with Naimark [85], Gelfand began to study Banach algebras in which there is an involution $x \mapsto x^*$ (i.e. such that $(x+y)^* = x^*+y^*$, $(xy)^* = y^*x^*$, $(\lambda x)^* = \bar{\lambda}x^*$ for any scalar λ and $(x^*)^* = x$) for which in addition $\|x^*x\| = = \|x\|^2$; these algebras are now called C^*-algebras. The main result proved by Gelfand and Naimark is that, for a commutative C^*-algebra A having a unit element e, the mapping $x \mapsto \mathcal{G}x$ is an isometry of A onto $\mathcal{C}(X(A))$ such that $\mathcal{G}x^* = \overline{\mathcal{G}x}$ for all $x \in A$. Furthermore, if there exists in A an element x_o such that the subalgebra of A generated by x_o, x_o^* and e is dense in A, then the map $\chi \mapsto \chi(x_o)$ is a homeomorphism of $X(A)$ onto $Sp_A(x_o)$, a compact subset of $\mathbb{C}$ which one therefore identifies with the spectrum of A.

The Gelfand-Naimark theorem paved the way for a new interpretation of Hilbert's spectral theory. Let E be a separable Hilbert space, N a continuous normal operator in E;
then the closure A in $\mathcal{L}(E)$ (for the normed topology) of
the algebra generated by 1_E, N and N^* is a separable
<u>commutative</u> C^*-<u>algebra</u> with unit, the mapping $\eta: \chi \mapsto \chi(N)$
being a homeomorphism of $X(A)$ onto the spectrum $Sp(N) \subset \mathbb{C}$.
From the Gelfand-Naimark theorem, it follows that the mapping
$f \mapsto \mathcal{G}^{-1}(f \circ \eta)$ is an <u>isometry</u> of the algebra $\mathcal{C}(Sp(N))$ on a
subalgebra of $\mathcal{L}(E)$, which one writes $f \mapsto f(N)$, obtaining
in this way a new definition of a "continuous function of a
normal operator" which had been considered by F. Riesz and
von Neumann. Following the method of von Neumann, it is then
easy to extend the homomorphism $f \mapsto f(N)$ to the algebra
$\mathcal{U}(Sp(N))$ of all <u>universally measurable</u> bounded functions in
$Sp(N)$.

Finally, by adapting the arguments of. von Neumann, Hellinger
and Hahn, one arrives at the modern description of the Riesz-
von Neumann "decomposition of unity" and of the "multiplicity
theory" of Hellinger-Hahn:

1º There is a decomposition of E into a Hilbert sum (fi-
nite or not) $(E_j)_{1 \le j < \omega}$ (ω being an integer or $+\infty$) of
closed subspaces, each of which is stable by N and by N^*.

2º There is a positive measure ν on the compact space
$Sp(N) \subset \mathbb{C}$ with support $Sp(N)$, and a decreasing sequence
$(S_j)_{1 \le j < \omega}$ with $S_1 = Sp(N)$, consisting of universally measurable sets.

3º For each j such that $1 \le j < \omega$, there is an <u>isometry</u>

T_j of the Hilbert space E_j onto the Hilbert space $F_j = L^2(\mathrm{Sp}(N),\varphi_{S_j}\cdot\nu)$ such that the normal operator $T_j\,N\,T_j^{-1}$ in F_j is the "multiplication operator" which, to the class of any function u_j defined and square integrable (for ν) in S_j, associates the class of the function $\zeta \mapsto \zeta u_j(\zeta)$.

4º In this description, the measure ν (considered as a measure on $\mathbb{C}$) is determined <u>up to equivalence</u>, the sets S_j are determined <u>up to a null set</u> (for ν). The set $M_j = S_j - S_{j+1}$ is the part of $\mathrm{Sp}(N)$ of <u>multiplicity</u> j, and (when $\omega = +\infty$), $M_\infty = \bigcap_j S_j$ the part of $\mathrm{Sp}(N)$ of <u>infinite multiplicity</u>. If P_j is the orthogonal projector $\varphi_{M_j}(N)$, and $H_{ik} = P_k(E_i)$, the restrictions of N to the k orthogonal subspaces H_{ik} ($1 \leq i \leq k$) are <u>equivalent</u>; the subspaces E_i are not uniquely determined, but the subspaces $G_k = P_k(E) = H_{1k} \oplus H_{2k} \oplus \ldots \oplus H_{kk}$ are. The <u>equivalence class</u> of ν and the <u>classes of the sets</u> S_j are the <u>unitary invariants</u> of N which determine it up to a unitary equivalence $N \mapsto UNU^{-1}$.

One says that this description is a <u>diagonalization</u> of the normal operator N. This name is justified when one considers the classical case in which N is a normal endomorphism of a finite dimensional space E: $\mathrm{Sp}(N)$ is then a discrete subset of $\mathbb{C}$ consisting of the eigenvalues of N, S_j the subset consisting of the eigenvalues of multiplicity $\geq j$, ν the measure having mass $+1$ at each point of $\mathrm{Sp}(N)$, and G_j is the subspace of E which is the direct sum of the eigenspaces of N corresponding to the eigenvalues of multiplicity j.

It is easy, using von Neumann's results, to extend the pre-

ceding description to <u>unbounded</u> normal operators N : $\mathrm{Sp}(N)$
is then an arbitrary closed subset of $\mathbb{C}$ (it may be $\mathbb{C}$ itself), and the S_j arbitrary universally measurable subsets
of $\mathrm{Sp}(N)$ forming a decreasing sequence; N is not defined
in the whole subspace E_j, but $\mathrm{dom}(N) \cap E_j$ is the subspace
transformed by T_j into the subspace of F_j consisting of
the u_j such that the function $\zeta \mapsto \zeta u_j(\zeta)$ is square integrable for ν.

Furthermore, for each universally measurable function f
in $\mathrm{Sp}(N)$ (bounded or not), $f(N)$ is a (generally unbounded) normal operator, which one may define in the following
way: $\mathrm{dom}(f(N)) \cap E_j$ is the subspace transformed by T_j into
the subspace of F_j consisting of the u_j such that the
function $\zeta \mapsto f(\zeta)u_j(\zeta)$ is square integrable, and the class
of this function is the image of the class of u_j by
$T_j f(N) T_j^{-1}$.

For self-adjoint operators A in E, the connection with
the "eigendifferentials" of Hellinger is made by the following remark, due to F. Riesz: for every $x = (x_k) \in \ell^2$, a
vector $(\rho_k(\xi))$ is defined by taking $A_\xi \cdot x$ for every $\xi \in \mathbb{R}$.

§6 - Later developments

Since 1940, an enormous number of papers have been published on Banach algebras, spectral theory and their applications.
I think a fair and well organized account of all these developments will have to wait till more time has elapsed and has

put them in their proper perspective[*]. With the exception
of the theory of differential (ordinary or partial) and integral equations, which has a complex background in which more
than spectral theory is involved, and which will be considered
in chap. IX, we shall limit our survey to bare indications of
the general trends, and to references to a few papers and
books.

A) Structure of Banach algebras.

After Gelfand and his school had investigated the general
properties of all Banach algebras, mathematicians concentrated their efforts on two particular classes of such algebras,
the commutative and the involutive ones.

For a commutative Banach algebra A, a central problem was
to define "functions" of elements $x \in A$ more general than
polynomials, after the pattern set by F. Riesz and von
Neumann. The latter had even shown that it was possible to
define functions $f(N_1, \ldots, N_k)$ of commuting normal operators
N_j in a Hilbert space, for all continuous functions f defined in C^k. For a general commutative Banach algebra A,
such a definition was only possible under some restrictions
on f; it $x_1, \ldots, x_k$ are any elements of A, their joint
spectrum is the image in $\mathbb{C}^k$ of the mapping

---

[*]Glaring examples of lack of perspective are given by the
Hellinger-Toeplitz article of 1923 in the Enzyklopädie der
math. Wiss. [107], which gives undue emphasis to integral
equations, and by Hadamard's article on Functional Analysis
of 1928 ([94], vol.I, p.435-453), which barely mentions
F. Riesz and does not speak of spectral theory at all!

$\chi \longmapsto (\chi(x_1), \chi(x_2), \ldots, \chi(x_k))$ where χ runs through $X(A)$
(for k = 1, it is of course the spectrum of x_1); one then
could prove that if B is the algebra of (germs of) functions
f <u>holomorphic in a neighbourhood</u> (<u>depending on</u> f) <u>of the</u>
<u>joint spectrum</u> of $x_1, \ldots, x_k$, there is a homomorphism B → A
written $f \longmapsto f(x_1, \ldots, x_k)$, which uniquely extends the natural
homomorphism of the algebra of polynomials on $\mathbb{C}^k$ into A
written similarly ([222], [27]).

On the spectrum $X(A)$ of a commutative Banach algebra A,
one soon was led to consider a topology different from the
one induced by the weak topology of the dual A′ of the
Banach space A. For Boolean rings, Stone had introduced the
idea of defining on the space of <u>maximal ideals</u> of such a
ring B a topology, in which the closed sets were defined as
the sets of maximal ideals containing a given (arbitrary)
ideal of B. As the set $X(A)$ of characters of a commutative
Banach algebra A corresponds in a one-to-one way to the set
of maximal ideals, Stone's topology can be defined in the same
way on $X(A)$; it is in general coarser than the weak topology,
and one says A is a <u>regular</u> commutative Banach algebra if
these two topologies on $X(A)$ coincide; for instance, the
algebra $\mathbb{C}(K)$ of continuous functions on a compact space K
is a regular algebra.

To any closed ideal J in A, one attaches the set $h(J)$
of all characters $\chi \in X(A)$ which vanish on J; a natural
question is to ask if the intersection of all kernels $\chi^{-1}(0)$
(maximal ideals of A) such that $\chi \in h(J)$, which always
contains J, is actually equal to J; one then says that

the ideal   J   admits   <u>spectral synthesis</u>.   Giving conditions
for a closed ideal to admit spectral synthesis in a regular
commutative Banach algebra is a problem which has been extensively studied ([21], [58]).

   <u>Involutive</u> Banach algebras   A   (not necessarily commutative)
are those equipped with an involution $x \mapsto x^*$ such that
$\|x^*\| = \|x\|$ for all $x \in A$; C^*-algebras (§5) are involutive
algebras, but there exist involutive Banach algebras which
are not C^*-algebras (see C) below).   The central concept is
that of   <u>representation</u> of an involutive Banach algebra   A   in
a Hilbert space   E;   this means a homomorphism $f: A \to \mathcal{L}(E)$
of algebras such that in addition $f(x^*) = f(x)^*$.   They have
been the subject of a large number of investigations, leading
to the elucidation of the structure of several classes of
C^*-algebras; the theory of von Neumann algebras (which are
special types of C^*-algebras) plays a great part in these investigations ([58], [36]).

B) <u>Algebras of continuous functions</u>.

   Since 1960, many mathematicians have been interested in
the study of <u>subalgebras</u> of Banach algebras $\mathcal{C}(K)$ of continuous functions on a compact space   K.   In classical
Analysis, one had much studied the case in which   K   is the
unit disk $|z| \leq 1$ in $\mathbb{C}$;   there is then in $\mathcal{C}(K)$ a parti-
cularly interesting Banach subalgebra, namely the algebra   B
of functions which are <u>holomorphic</u> in the interior $|z| < 1$
of the disc.   It can be identified with the algebra B_0 of
the <u>restrictions</u> of the functions of   B   to the unit circle $\mathbb{U}$:
$|z| = 1$,   and B_0 is also the closure in $\mathcal{C}(\mathbb{U})$ of the alge-

bra of _trigonometric polynomials_. It turns out that the
study of B_o is closely linked to the _completions_ of the
space of trigonometric polynomials in the various spaces $L^p(\mu)$,
where μ is Haar measure on $\mathbb{U}$, and many beautiful properties of these spaces (known as the _Hardy spaces_ $H^p(\mu)$) had
been discovered. But in the light of the theory of commutative Banach algebras, it was found that these results could
be much better understood if they were generalized to subalgebras of an algebra $C(K)$ where K is any compact space,
and put in relation with some kinds of measures on K (see
[18], [33], [79], [116], [140]).

C) _Harmonic Analysis_.

We have already stressed the fact (chap.I, §2) that
Fourier series provided the starting point of spectral theory
when it was realized that they could be generalized to "expansions" in series of "orthogonal" functions arising from
boundary value problems.

It was, however, very soon observed that the "trigonometric
system" $(e^{inx})_{n \in \mathbb{Z}}$ possessed very peculiar properties not
shared by general "orthogonal systems", and linked to the
functional equation $e^{i(x+x')} = e^{ix}e^{ix'}$. For instance, if
$f(x) = \sum_{n=-\infty}^{+\infty} a(n)e^{nix}$, $g(x) = \sum_{n=-\infty}^{+\infty} b(n)e^{nix}$ were two Fourier
series, one had for the Fourier series $f(x)g(x) =$
$= \sum_{n=-\infty}^{+\infty} c(n)e^{nix}$ of their product, the very simple formula

$$(40) \qquad\qquad c(n) = \sum_{p=-\infty}^{+\infty} a(p)b(n-p).$$

Similarly, from the formula $f(x) = \sum_{n=-\infty}^{+\infty} a(n)e^{nix}$, one obtained

$$(41) \qquad \sum_{n=-\infty}^{+\infty} a(n+1)e^{nix} = e^{-ix}f(x)$$

and this property was used by de Moivre and even more by

Laplace to solve linear difference equations $\sum_{k=1}^{p} \alpha_k a(n+k) = 0$

by associating to the sequence $(a(n))$ the Fourier series

$f(x) = \sum_{n} a(n)e^{nix}$, reducing the difference equation to an

algebraic equation for $f(x)$.

Similar peculiarities were observed for the "Fourier transform" associating to an integrable function f in $\mathbb{R}$ the
function

$$(42) \qquad \mathfrak{F}f(x) = \int_{-\infty}^{+\infty} e^{-2\pi ixt} f(t)dt.$$

Its main virtue, in the eyes of Fourier, Cauchy and Poisson,
was that it reduced linear partial differential equations with
constant coefficients to algebraic problems, due to the fact
that the Fourier transform of the derivative f' is the
function $x \mapsto 2\pi ix\, \mathfrak{F}f(x)$. Furthermore, in his researches on
Probability theory, Tchebycheff had shown that if F_1, F_2 are
two independent "random variables" with "probability laws"
α_1, α_2, measures on $\mathbb{R}$ with densities g_1, g_2, the probability law of $F_1 + F_2$ had a density given by the convolution
$g = g_1 * g_2$, defined as

$$(43) \qquad g(x) = \int_{-\infty}^{+\infty} g_1(t)g_2(x-t)dt \qquad ([211], vol.II, p.481-491);$$

and Tchebycheff's student Liapounov, who started to use Fourier
transforms in Probability theory, observed that

$$(44) \qquad \mathfrak{F}(g_1 * g_2) = \mathfrak{F}g_1 \cdot \mathfrak{F}g_2 \qquad [148].$$

One should also mention the Poisson formula (also discovered independently by Cauchy)

$$(45) \qquad\qquad \sum_{n \in \mathbb{Z}} f(n) = \sum_{n \in \mathbb{Z}} \mathfrak{F}f(n)$$

for sufficiently regular functions f on $\mathbb{R}$.

It took over 100 years to understand these peculiarities
and to connect them with the notion of group, via the concepts
of character and of group algebra. Characters were first defined for arbitrary finite commutative groups by H. Weber in
1882, as complex valued functions χ on such a group G with
values $\neq 0$, such that $\chi(xy) = \chi(x)\chi(y)$ for all x, y in G;
but special cases had long before been considered by Legendre,
Gauss and Dirichlet. A meaningful generalization to non commutative finite groups was discovered in 1896 by Frobenius:
instead of considering homomorphisms of G into the multiplicative group $\mathbb{C}^*$, one should consider homomorphisms
$s \longmapsto U(s)$ of G into the general linear group $GL(n,\mathbb{C})$ of
invertible matrices of order n., for any integer n. This
is also called a linear representation of degree n of G
in the vector space $E = \mathbb{C}^n$: giving such a representation is
equivalent to defining an action $(s,x) \longmapsto s \cdot x$ of G on E
such that $s \cdot (t \cdot x) = (st) \cdot x$, $e \cdot x = x$ for the neutral element
$e \in G$, and such that each mapping $x \longmapsto s \cdot x$ is linear (with
matrix $U(s)$).

Now Cayley had defined, for a finite group G, the group
algebra $\mathbb{C}[G]$ as the vector space of all formal linear combinations $\sum_{s \in G} \xi_s s$ with $\xi_s \in \mathbb{C}$, multiplication being defined by

(46) $(\sum_{s\in G} \xi_s s)(\sum_{s\in G} \eta_s s) = \sum_{(s,t)\in G\times G} \xi_s\eta_t st.$

When there is given an action $(s,x) \mapsto s\cdot x$ of G on E as above, it defines naturally on E a structure of <u>left</u> $\mathbb{C}[G]$-<u>module</u> by

(47) $(\sum_{s\in G} \xi_s s)\cdot x = \sum_{s\in G} \xi_s(s\cdot x)$

and the study of linear representations of G is thus equivalent to the study of left $\mathbb{C}[G]$-modules.

The fundamental results of Frobenius for finite groups may then be expressed in the following way. An element $\sum_{s\in G} \xi_s s$ of $\mathbb{C}[G]$ may be identified with a mapping $f: s\mapsto \xi_s$ of G into $\mathbb{C}$, so that $\mathbb{C}[G]$ may be identified with the vector space $\mathbb{C}^G$ of <u>all</u> mapping of G into $\mathbb{C}$, with the multiplication written $f*g$ and defined by

(48) $(f*g)(s) = (\mathrm{Card}(G))^{-1} \sum_{t\in G} f(t)g(t^{-1}s).$

Then $\mathbb{C}[G]$ can be written as a direct sum $A_1 \oplus A_2 \oplus\ldots\oplus A_h$ of mutually annihilating subalgebras, where h is the number of classes of conjugate elements in G; each A_k $(1\leq k\leq h)$ is a <u>matrix algebra</u> of dimension n_k^2 over $\mathbb{C}$, which means that it has a basis (m_{ij}^k) of n_k^2 elements belonging to $\mathbb{C}^G$, with the following properties:

(49) $m_{pq}^k * m_{rs}^k = \delta_{qr} m_{ps}^k$ for $1 \leq p,q,r,s \leq n_k$

(50) $m_{pq}^k * m_{rs}^{k'} = 0$ if $k \neq k'.$

In addition, one has

(51) $m_{ji}^{k}(s) = \overline{m_{ij}^{k}(s^{-1})}$ for $1 \leq i,j \leq n_k$, $s \in G$

(52) $\sum_{s \in G} m_{pq}^{k}(s)m_{rs}^{k'}(s) = 0$ unless $p=r$, $q=s$, $k=k'$

(53) $\sum_{s \in G} m_{pq}^{k}(s)\overline{m_{pq}^{k}(s)} = n_k$ Card(G) for $1 \leq p,q \leq n_k$

(orthogonality relations). One has $n_1^2 + n_2^2 + \ldots + n_h^2 =$ Card(G)
and the expression of any $f \in \mathbb{C}^G$ with respect to the basis
(m_{ij}^{k}) of that space,

(54) $f(s) = \sum_{i,j,k} c_{ij}^{k} m_{ij}^{k}(s)$

is given explicitly, due to the orthogonality relations, by

(55) $c_{ij}^{k} = (n_k$ Card$(G))^{-1} \sum_{s \in G} f(s)\overline{m_{ij}^{k}(s)}$.

If one writes $M_k(s)$ the $n_k \times n_k$ matrix $(n_k^{-1}m_{ij}^{k}(s))$,   one
has

(56) $M_k(st) = M_k(s)M_k(t)$ and $M_k(s^{-1}) = (M_k(s))^*$

In other words, $s \mapsto M_k(s)$ is a linear representation of   G
of degree n_k for $1 \leq k \leq h$;   it is irreducible, which means
that the corresponding $\mathbb{C}[G]$-module is simple (i.e. has no
non trivial submodule). Furthermore, every $\mathbb{C}[G]$-module of
finite dimension over $\mathbb{C}$ is a direct sum of modules each of
which corresponds to one of the linear representations
$s \mapsto M_k(s)$;   one says that every linear representation of   G
is completely reducible, and that it contains the irreducible
representation $s \mapsto M_k(s)$ with multiplicity d_k if in the
direct decomposition of the corresponding module, there are
d_k submodules corresponding to $s \mapsto M_k(s)$.

Now linear representations of degree n can be defined in
the same way for any group G, finite or not. Already in
1901, I. Schur, in his dissertation ([193],vol.I, p.1-70),
could determine all linear representations of the general linear group GL(N,ℂ) which are such that the elements of $U(s)$
are polynomials in the elements of the matrix s ∈ GL(N,ℂ);
he showed that these representations are again completely reducible and he could determine explicitly the irreducible ones
(see [53]); but it is clear that for such infinite groups, all
the Frobenius relations described above were meaningless.
However, in 1924, I. Schur observed that the restrictions of
these representations to the group of rotations G = SO(N,ℝ)
gave him irreducible representations of that compact group,
and that these representations were continuous, and could be
written s ⟼ $M_k(s)$ where $M_k(s)$, as in (56), was a unitary
matrix; furthermore, he proved the relations which he rightly
considered as the analogues of (52)

$$(57) \qquad \int_G m^k_{pq}(s)\overline{m^{k'}_{rs}(s)}\,ds = 0 \quad \text{unless} \quad p=r, \; q=s, \; k=k'$$

where   ds is a left and right invariant measure on G, the
existence of which was substantially known since S. Lie, and
which had already been used to construct invariants of SO(N,ℝ)
by Hurwitz in 1898 ([193], vol.II, p.440-494).

This result attracted the attention of H. Weyl; in a beautiful series of 3 papers published the next year, by a skill-
ful combination of Schur's ideas with the "infinitesimal"
methods by which E. Cartan in 1913 had obtained all finite
dimensional representations of the complex semi-simple Lie

groups, he was able to determine explicitly <u>all</u> continuous irreducible linear representations of <u>compact semi-simple Lie groups</u> (including, in the case of $SO(N,\mathbb{R})$, the "spinor" representations which had escaped I. Schur). In all cases, the orthogonality relations (57) still held, and every continuous linear representation of a semi-simple compact Lie group was shown to be completely reducible ([227],vol. II, p. 633).

Of course, for compact Lie groups, there is an <u>infinite</u> system of irreducible representations M_k, and relations such as (54) were out of the question. But for the group $SO(1,\mathbb{R})=$ $= \mathbb{U}$ (the circle group), the irreducible representations were the characters $\zeta \mapsto \zeta^n$ for $n \in \mathbb{Z}$, and H. Weyl realized that the formula which corresponded to (54) was just the <u>Fourier series expansion</u> of f (when f is sufficiently regular) [227, vol.III, p.34-37]. He then undertook to generalize this expansion to all semi-simple compact Lie groups G; for such a group, the functions m_{ij}^k which he had determined formed an orthogonal system in the Hilbert space $L^2(G)$ (for a left and right invariant measure); the problem was to prove that this system was <u>complete</u>.

This is what H. Weyl proved in 1927, in a remarkable paper written in collaboration with his student F. Peter [227, vol. III, p.58-75], which can be considered as the first application of spectral theory to harmonic analysis. He saw that the notion which could serve as a substitute to the group algebra $\mathbb{C}[G]$ was the space $\mathbb{C}(G)$ of continuous complex-valued functions on the compact group G, on which an algebra structure is defined by <u>convolution</u>, generalizing (40), (43), and (48):

(58) $$(f*g)(s) = \int_G f(t)g(t^{-1}s)dt$$

where the integration is for a left and right invariant positive measure on G with total mass 1. H. Weyl next observed that, given a linear representation $s \longmapsto U(s)$ of G by unitary matrices of order n, if one wrote

(59) $$U(f) = \int_G f(s)\overline{U(s)}ds$$

one obtained a <u>homomorphism</u> of the algebra $C(G)$ into $\text{End}(\mathbb{C}^n)$, in other words

(60) $$U(f*g) = U(f)U(g)$$

and furthermore, if one wrote $\overset{\vee}{f}(s) = \overline{f(s^{-1})}$, one had

(61) $$U(\overset{\vee}{f}) = (U(f))^*.$$

The crux of his proof is to show that for an $f \neq 0$ in $C(G)$, there is at least one continuous representation $s \mapsto U(s)$ for which $U(f) \neq Q$; due to the complete reducibility of $s \longmapsto U(s)$, there is then at least one irreducible representation $s \mapsto M_k(s)$ such that $M_k(f) \neq 0$, and this shows that f cannot be orthogonal to all functions m_{ij}^k. However, if the system (m_{ij}^k) was not complete, there would exist a non negligible function $g \in L^2(G)$ orthogonal to all the m_{ij}^k, and as the subspace L of $C(G)$ generated by the m_{ij}^k is a two-sided ideal (one has $(f*U)(s) = U^*(f)U(s)$), $h*g$ would also be orthogonal to L for any function $h \in C(G)$, and one has $h*g \in C(G)$ and $h*g \neq 0$ for suitable functions h. [*]

[*]This is a slight simplification of Weyl's argument, which consists in obtaining for each function of $C(G)$ the analogue of the Fischer-Riesz expansion by an inductive application of the Schwarz-E.Schmidt method.

The proof of the existence of a representation U such that $U(f) \neq 0$ is deduced by Weyl from the theory of Hilbert-Schmidt integral equations. He considers the function $g = f * \overset{\vee}{f}$, for which $U(g) = U(f)U^*(f)$, and it is enough to show that $U(g) \neq 0$ for some U. One has $g(e) = \int_G |f(t)|^2 \, dt > 0$; Weyl forms the sequence of functions $g_1 = g$, $g_2 = g * g_1, \ldots$, $g_n = g * g_{n-1}, \ldots$; by an adaptation of the method of H.A. Schwarz, as generalized by E. Schmidt to integral equations with symmetric kernels (chap. III, §1 and chap. V, §2), he proves that the sequence of numbers $\gamma_n = g_n(e)/g_{n-1}(e)$ is increasing and tends to a limit $\gamma > 0$, and g_n/γ_n tends uniformly to a continuous function u, such that $g * u = u * g = \gamma u$ and $u * \overset{\vee}{u} = u$; γ is an eigenvalue of the hermitian kernel $k(s,t) = g(st^{-1})$, and if φ_j $(1 \leq j \leq r)$ form an orthonormal basis of the corresponding eigenspace, one easily proves that $u(st^{-1}) = \varphi_1(s)\overline{\varphi_1(t)} + \ldots + \varphi_r(s)\overline{\varphi_r(t)}$. Furthermore, for any $t \in G$, the function $s \mapsto \varphi_j(st^{-1})$ again is an eigenvector of the same space, hence $\varphi_j(st^{-1}) = \sum_{k=1}^{r} \overline{u_{ik}(t)}\varphi_k(s)$, and one shows that if $U(t) = (u_{jk}(t))$, $t \mapsto U(t)$ is a linear representation of G for which $U(g) \neq 0$.

H. Weyl himself remarked that this method also proved the existence of the irreducible representations M_k, and was applicable to any compact Lie group (not necessarily semi-simple); a little later, when A. Haar had proved in 1933 the existence of a measure invariant by left and right translation on **any** compact subgroup, Weyl's arguments could at once be extended to that general case.

The next year, Pontrjagin, in view of applications to alge-

braic topology, showed that the Peter-Weyl theory, applied to
commutative metrizable compact groups, led to a remarkable
generalization of the duality between finite commutative
groups, which had been well-known since Weber. It was of
course classical that an irreducible linear representation of
a commutative group G must be of degree 1, in other words
it is a character χ of G. The Peter-Weyl theory therefore
associated to a metrizable compact commutative group G the
set $\hat{G}$ of all continuous characters of G, which of course
is itself a denumerable group for ordinary multiplication.
Now, for any x ∈ G, the map χ ↦ χ(x) is clearly a character of $\hat{G}$, and Pontrjagin showed that all characters of $\hat{G}$
are of that type. Conversely, if D is any denumerable group,
and $\hat{D}$ the group of all its characters, one can put on $\hat{D}$
the topology of simple convergence, for which it becomes a
metrizable compact group, and then all continuous characters
of $\hat{D}$ are exactly the maps χ ↦ χ(x) for all x ∈ D.

But Pontrjagin went further and could extend this duality
to some locally compact commutative groups [179], and in 1936
van Kampen showed, by different methods, that Pontrjagin's
results could be generalized to all such groups G. The dual
$\hat{G}$, consisting of all continuous characters on G, is given
the topology of uniform convergence on compact subsets of G,
and it is again locally compact for that topology; to each
x ∈ G there corresponds the continuous character η(x):
χ ↦ χ(x) on $\hat{G}$, and the duality theorem of Pontrjagin-van
Kampen says that η is an isomorphism of topological groups
of G onto $\hat{\hat{G}}$ [216].

This discovery made possible a unified treatment of the Fourier series and of the Fourier integral. In general, for any function $f \in L^1(G)$, one could define the function on $\hat{G}$

$$(62) \qquad \mathfrak{F}f(\chi) = \int_G f(x)\overline{\chi(x)}dx$$

as the Fourier transform of f. For G = $\mathbb{R}$, continuous characters could be uniquely written $x \longmapsto \exp(2\pi ixy)$ for a real number y, so that $\hat{G}$ could be identified with $\mathbb{R}$ itself, and (62) was just the definition of the Fourier integral. For G = $\mathbb{Z}$, all characters are continuous and can be written $n \mapsto \zeta^n$ for a uniquely determined complex number $\zeta \in U$; for a function $n \mapsto c(n)$ on $\mathbb{Z}$ such that $\sum_n |c(n)| < +\infty$, the ·right hand side of (62) becomes the absolutely convergent Fourier series $\sum_{n \in \mathbb{Z}} c(n)\zeta^n = \sum_{n \in \mathbb{Z}} c(n)e^{ni\theta}$ if $\zeta = e^{i\theta}$, a function defined on the dual U of $\mathbb{Z}$. Finally, for G = U, continuous characters are the functions $\zeta \mapsto \zeta^n$ for a uniquely determined $n \in \mathbb{Z}$, and the Fourier transform of a function $f \in L^1(U)$ is the sequence $n \mapsto c(n)$ of its "Fourier coefficients". In 1940, A. Weil [226] showed how most results concerning Fourier series and integrals could be generalized to all locally compact commutative groups; the central theorem was the generalization of the Parseval relation: if a function f on G belongs to $L^1(G) \cap L^2(G)$, its Fourier transform belongs to $L^2(\hat{G})$, and one has the relation

$$(63) \qquad \int_G |f(x)|^2 \, dx = \int_{\hat{G}} |\mathfrak{F}f(\chi)|^2 \, d\chi$$

for a suitable Haar measure on $\hat{G}$; this relation, for the case G = $\mathbb{R}$, had been proved in 1910 by M. Plancherel [174],

and is known as the Plancherel theorem for locally compact
commutative groups.

We have already mentioned (§5) that Gelfand, when he defined
characters on a commutative Banach algebra, had followed the
pattern set by H. Weyl and Pontrjagin. In fact, in a joint
paper with D. Raikov [86], he immediately showed how the
Pontrjagin-van Kampen-A. Weil theory could be deduced from his
general results on Banach algebras in a much simpler way (the
earlier proofs relied heavily on detailed information on the
structure of locally compact commutative groups). The basic
idea is to consider, for a locally compact commutative group
G, the space $L^1(G)$ (for a Haar measure on G), on which a
structure of Banach algebra is defined by the convolution
product (58); it is an involutive algebra for the involution
$f \mapsto \overset{\vee}{f}$, but in general it is not a C^*-algebra. A character
of that algebra (in the sense of Gelfand) can then be unique-
ly written as $f \mapsto \mathfrak{F}f(\chi)$ for a well-determined character χ
(in the sense of Pontrjagin), so that the spectrum $X(L^1(G))$
is identified (with its topology induced by the weak topology
of $L^\infty(G)$) with the dual group $\hat{G}$, and then the Fourier
transform merely becomes a special case of the Gelfand transform, equation (44) being the expression in that special case
of the fact that the Gelfand transform is a homomorphism of
algebras!

But this absorption of harmonic Analysis by spectral theory
did not stop with commutative groups. One can still define
the Banach algebra $L^1(G)$ for locally compact separable uni-

modular groups (i.e. those for which left invariant Haar measure is also right invariant, for instance compact groups or
semi-simple Lie groups), and it is still an involutive algebra.
On the other hand, one can define linear representations
$s \mapsto U(s)$ of such a group G not only when $U(s)$ is a unitary matrix, but more generally when $U(s)$ is an automorphism
of a complex Hilbert space E; one then speaks of unitary representations of G in E, and one adds to the definition
the additional condition that for any $x \in E$, the mapping
$s \mapsto U(s) \cdot x$ of G into the Hilbert space E should be continuous. For any function $f \in L^1(G)$, it is then possible
to define $U(f)$ as in (59), more precisely, one has, for any
$x \in E$ and $y \in E$

$$(64) \qquad (U(f) \cdot x \mid y) = \int_G f(s)(U(s) \cdot x \mid y) ds$$

where ds is (left and right) invariant Haar measure on G.
It is then remarkable that starting from the unitary representations of G in E, one obtains in this way a bijection
of the set of these representations onto the set of all homomorphisms V of $L^1(G)$ into the C^*-algebra $\mathcal{L}(E)$ which satisfy (61) and are non-degenerate (i.e. such that the $V(f) \cdot x$
for $x \in E$ and $f \in L^1(G)$ generate a dense subspace of E).
With convenient modifications, there is still a similar result for all locally compact groups, and the general theory
of unitary representations of locally compact groups in
Hilbert space is thus in a certain sense subordinate to the
theory of homomorphisms of involutive Banach algebras in algebras of operators in Hilbert space [58].

However, most results concerning unitary representations of locally compact groups have up to now been restricted to Lie groups, where a large number of more refined and powerful tools (Lie algebras, differential geometry, partial differential equations, etc.) are available. We can only mention here this beautiful and difficult theory (known as non commutative harmonic Analysis), which has known an enormous expansion since 1950, and in which many problems are still open; the interested reader is referred to [32], [43], [152], [217] and [223]; for a detailed history of harmonic Analysis (both commutative and non commutative) and its relations with probability theory, quantum mechanics and number theory, see [155].

D) Other developments.

One of the first results on infinite dimensional representations was obtained by M.H. Stone in 1930 [206]: he showed that any unitary representation of the additive group $\mathbb{R}$ into a separable Hilbert space E was given by the formula $t \mapsto e^{itA}$, where A is an arbitrary (in general unbounded) self-adjoint operator in E, so that one may say that the theory of unitary representations of $\mathbb{R}$ is equivalent to the spectral theory of unbounded self-adjoint operators.

If now E is an arbitrary Banach space, and A a bounded operator in E it is clear that $t \mapsto e^{tA}$ is a homomorphism of $\mathbb{R}$ into the group of invertible elements in $\mathcal{L}(E)$. Unbounded operators A may be defined in E just as in Hilbert space, but for such an operator, e^{tA} usually has no meaning for every real number t. Various questions of Analysis led E. Hille, in a series of papers beginning in 1936, to in-

vestigate mappings $t \mapsto P_t$ into $\mathcal{L}(E)$, <u>only defined for</u>
$t > 0$ and such that, for $s > 0$ and $t > 0$

(65) $P_{s+t} = P_s P_t .$

Such mappings are called <u>semi-groups of operators</u>. If one
assumes that $\|P_t\| \leq C$ for all $t > 0$ and that for every
$x \in E$, $t \mapsto P_t \cdot x$ is continuous for $t > 0$, then one may
associate to such a semi-group an <u>unbounded operator</u> A in E,
defined by

(66) $A \cdot x = \lim_{t \to 0} t^{-1}(P_t - 1_E) \cdot x .$

This has been the starting point of an extensive theory with
many applications in Analysis [115].

A large literature has been devoted to various types of operators in Banach spaces. A very general method consists in
starting with an operator A_o whose properties are well-known
(for instance a normal operator in a Hilbert space) and to
consider operators $A = A_o + P$ which differ from A_o by a
"perturbation" P which is "small" in some sense; for instan-
ce, the norm $\|P\|$ is supposed to be small enough, or P is a
<u>compact</u> operator; such assumptions allow in many cases to
extend some properties of A_o to A (see [124]).

The nice properties of the operators $1_E + K$, where K is
a compact operator (§1) have inspired the study of generalizations of such operators, for instance the <u>Fredholm operators</u>
U, which are defined by the properties that $U^{-1}(0)$ has finite dimension, $U(E)$ is closed and has finite codimension,
but the dimension of $U^{-1}(0)$ and the codimension of $U(E)$
are not necessarily equal [133]. Finally, there is an exten-

sive theory of operators for which there is a family of projectors having properties similar to the projectors $E(\lambda)$
associated to a self-adjoint operator in von Neumann's theory
(§4); the difficulty is of course to find criteria implying
the existence of such a family ([62], vol.III).

CHAPTER VIII

LOCALLY CONVEX SPACES AND THE THEORY OF DISTRIBUTIONS

§1 - Weak convergence and weak topology

In his thesis, Fréchet had already noticed that convergence
in a metric space could not always correspond to some classical types of "convergence" for functions. For instance, if
$\mathfrak{B}(\mathbb{R})$ is the vector space of <u>all</u> bounded real functions on $\mathbb{R}$,
it is not possible to define a distance on that space such
that <u>simple</u> convergence in $\mathfrak{B}(\mathbb{R})$ would be identical with convergence for that distance. This results from the fact that
if A is a subset of a metric space E, the closure $\bar{A}$ of
A in E is identical to the set of limits of all convergent
sequences of elements of A. However, if one takes in $\mathfrak{B}(\mathbb{R})$
the set $A = C(\mathbb{R})$ of bounded continuous functions, the limits
of sequences of elements of A for simple convergence are
the Baire functions of class 1, and it is known that there
are Baire functions of class 2 which are not of class 1, so
that $\bar{A}$ (for the hypothetical distance) could not consist
only of functions of class 1 [71, p.15].

There was thus an obvious need for a generalization of the
concept of metric space, but none proved adequate for Functional Analysis until Hausdorff, in 1914, created "General topology" as we understand it now, based on the concept of neigh-

borhood [100] ; but surprizingly enough, it took some time to become aware of that adequacy. Ever since Hilbert, "weak convergence" of <u>sequences</u> had become a central theme, first in Hilbert spaces, then with F. Riesz and Helly in some types of normed spaces (chap.VI), and one would have thought that Hausdorff's concept of topology would have been tested on that notion; but until 1934 the only mathematician who seems to have had that idea was von Neumann: he defined weak neighborhoods of a point x_o in a Hilbert space E by a finite number of conditions $|(x-x_o|a_j)| \leq \varepsilon$ for points $a_j \in E$, and then went on to define similarly, in the algebra $\mathcal{L}(E)$ of endomorphisms of E, "strong neighborhoods" of an operator U_o by a finite number of conditions $\|(U-U_o)\cdot x_j\| \leq \varepsilon$ for $x_j \in E$, and "weak neighborhoods" by a finite number of conditions $|((U-U_o)\cdot x_j|y_j)| \leq \varepsilon$ [221, vol.II, p.94-104]. But he did not try to extend these ideas to other Banach spaces.

On the contrary, "weak convergence" was at the center of Banach's book, and the results he obtained concerning that notion can be considered as some of his deepest work. But to understand what he did, it is probably better first to state the final form which was taken by his 3 main theorems:

I) If E is a Banach space, and its dual E' is given the weak topology $\sigma(E',E)$, the unit ball $\|x'\| \leq 1$ in E' is compact for that topology.

II) In order that a vector subspace $V \subset E'$ be closed for the topology $\sigma(E',E)$, it is necessary and sufficient that for any closed ball B' in E', $V \cap B'$ be compact for that topology.

III) In order that E be reflexive, it is necessary and
sufficient that the unit ball $\|x\| \leq 1$ in E be compact for
the weak topology $\sigma(E,E')$.

In this form, the theorems were proved independently by
L. Alaoglu ([5], [5 bis]) and N. Bourbaki [25] in 1938. Their
proofs use the following ingredients:

a) The weak topology $\sigma(E',E)$ is defined by taking as neighborhoods of $x'_0 \in E'$ the sets defined by a finite number of
relations $|\langle x'-x'_0, x_j \rangle| \leq \varepsilon$ for arbitrary $x_j \in E$; $\sigma(E,E')$
is defined similarly by exchanging the roles of E and E'.

b) The word "compact" is used in the sense of N. Bourbaki,
and means what was defined as "bicompact sets" (in Hausdorff
spaces) by P. Alexandroff and P. Urysohn in 1924 [6]:   for
them a space is "bicompact" if every open covering of the
space contains a finite covering (the "Borel-Lebesgue axiom").

c) Compact sets can be characterized equivalently by means
of the notion of limit of a "net", a notion which generalizes
the limit of a sequence and was introduced in 1922 by E.H.
Moore and H.L. Smith [164]   (N. Bourbaki uses the equivalent
concept of limit of a "filter").

d) Any product of compact spaces is compact, a theorem
proved by A. Tychonoff in 1930.

However, none of these notions or theorems was ever mentioned by Banach or mathematicians of his school until 1940,
although they repeatedly quote Hausdorff's book of 1927 [*];

---

[*] The bulk of that book [101] is devoted to metric spaces,
and general topological spaces are given a very scanty treatment in 5 pages; it seems that Hausdorff had lost faith in
his ideas of 1914!

for them, the word "compact" is always taken in the initial
sense of Fréchet, meaning a space in which there is no infinite closed discrete set. That notion is equivalent to the
notion of "bicompact" space when restricted to metrizable
spaces; the version of theorem I proved by Banach [15,p.123]
is therefore limited to separable Banach spaces E (because
in that case the ball $\|x'\| \leq 1$ is metrizable for the weak
topology $\sigma(E',E)$). It generalizes of course the "principles
of choice" used by Hilbert, F. Riesz and Helly (chap.V and VI).

On the other hand, Banach was able to prove a theorem equivalent to Theorem II for all Banach spaces. He starts from
the study of vector subspaces of E' which are closed for
the topology of the norm, and observes that for such a subspace V, a sequence (x'_n) of points of V may have a weak
limit which does not belong to V; furthermore, if V_1 is
the vector space consisting of all these weak limits, it may
happen that sequences of points of V_1 have weak limits
which do not belong to V_1, and so on [15,p.209]. Without
speaking of weak topology, he then introduces the weakly
closed vector subspaces, under the name of "regularly closed"
subspaces: he defines such a space V by the property that,
for any $x'_0 \notin V$, there is an $x \in E$ such that $\langle x',x \rangle = 0$
for all $x' \in V$, but $\langle x'_0,x \rangle \neq 0$. Conscious of the fact that
weakly convergent sequences are inadequate tools, Banach then
introduces an ad hoc notion, the "limits of bounded transfi-
nite sequences" in the dual E': for any family $(u'_\xi)_{\xi < \gamma}$ of
elements of E', contained in a ball and indexed by a segment
of the ordinals, he shows (by using the Hahn-Banach theorem)

that there always exist elements $u' \in E'$ such that, for

every $x \in E$,

$$\liminf_{\xi \to \gamma} \langle u'_\xi, x \rangle \leq \langle u', x \rangle \leq \limsup_{\xi \to \gamma} \langle u'_\xi, x \rangle$$

and calls any such u' "a limit" of the transfinite sequence

(u'_ξ) (there may be infinitely many such "limits"!). He then

says V is "<u>transfinitely closed</u>" if every bounded transfini-

te sequence of elements of V has at least a "transfinite

limit" in V, and what he shows, by a very clever argument,

is that "regularly closed" and "transfinitely closed" are

<u>equivalent</u> notions [15, p.121]. It was an easy matter to

replace "transfinite sequences" by "nets" or "filters" in

Banach's proof to obtain the equivalent statement of Theorem

II.

Finally, by considering E as naturally imbedded in its

second dual E'' and using again "transfinite limits", Banach

could prove Theorem III, but only when E is <u>separable</u> (*).

It should be mentioned here that these theorems enabled

Banach to obtain a series of interesting theorems relating the

properties of a continuous linear mapping $u: E \to F$ (where E

and F are Banach spaces), those of its transposed mapping

$^t u: F' \to E'$, and properties of the images $u(E)$ and $^t u(F')$

(see [51]) (**).

(*) In 1938, Goldstine proved a result which is equivalent to
the property that for any Banach space **E**, the intersection
$E \cap B''$, where B'' is the unit ball $\|x''\| \leq 1$ in the second dual E'',
is dense in B'' for the weak topology $\sigma(E'', E')$; from this
Theorem III easily follows [88].

(**) Some of Banach's results had been obtained by Hausdorff in
1931 [102]; it is quite remarkable that he makes no mention of
weak topology!

§2 - Locally convex vector spaces

Although the theory of normed spaces was in the forefront
of the development of Functional Analysis after 1906, it was
soon realized that they did not exhaust the possibility of
applying topological concepts to that discipline; but the various notions belonging to what we now call the general theory
of <u>topological vector spaces</u> made their appearance in a rather
random way and were not the subject of a systematic treatment
until 1950.

Already some examples of such spaces are to be found in
Fréchet's thesis, where the emphasis is put, not on their algebraic properties but on the possibility of defining their
topology by a distance (chap.V, §3) and on the fact that the
metric spaces thus obtained are complete. The fact that addition and multiplication by a scalar are continuous in such
spaces was only explicitly emphasized by Fréchet in 1926 [73];
the idea was picked up by Banach who in his book considered
in general these spaces under the name of "spaces of type (F)"
and showed that the closed graph theorem was also valid for
them. A little later, the method which Fréchet had used to
define the distance on his examples of 1906 was systematized
by S. Mazur and W. Orlicz in what they called the theory of
"spaces of type (B_o)" [160]:   they are what we now call
<u>Fréchet spaces</u>, where the topology is defined by a <u>sequence</u>
(p_n) of seminorms with the condition that $x \neq 0$ implies
that $p_n(x) \neq 0$ for at least one index n;   the distance can
be defined by the formula

$$d(x,y) = \frac{p_1(x-y)}{1+p_1(x-y)} + \frac{1}{2!} \frac{p_2(x-y)}{1+p_2(x-y)} + \cdots + \frac{1}{n!} \frac{p_n(x-y)}{1+p_n(x-y)} + \cdots$$

and it is supposed that the space is complete for that distance. For the examples of Fréchet (the space $\mathbb{R}^{\mathbb{N}}$ of all sequences and the space of holomorphic functions in $|z| < 1$) it can easily be shown that the topology cannot be defined by a single norm.

Other types of spaces were not even metrizable; this was observed by von Neumann in 1929 for the weak topology on Hilbert space (§1). But already in 1910 E.H. Moore had put forward the idea of replacing uniform convergence in $\mathbb{R}$ by what he called "relative uniform convergence"; this amounts to consider neighborhoods of 0 defined in the following way: one considers continuous functions g in R such that $g(x) > 0$ for all $x \in \mathbb{R}$, and to each such function g, one associates a neighborhood V_g of 0 consisting of all functions f such that $|f| \leq g$; when restricted to functions which are continuous and have compact support, these neighborhoods are exactly those defining what will later be called the (LF)-topology on $\mathcal{K}(\mathbb{R})$ [163].

After 1932, a new notion emerged, that of boundedness. It was already realized by Banach that on the same vector space, two norms $\|x\|_1$ and $\|x\|_2$ such that the ratios $\|x\|_1/\|x\|_2$ and $\|x\|_2/\|x\|_1$ are bounded for $x \neq 0$, define the same topology and therefore, if one defined a bounded set in a normed space E as being contained in some ball, this was a notion independent of the particular norm chosen. However, in an arbitrary metric space, two distances may give rise to the same

topology and give quite different notions of "bounded sets"
when one sticks to the previous definition. But for arbitrary
topological vector spaces (i.e. those for which addition and
multiplication by a scalar are continuous), it turned out that
it is possible to give a definition of bounded set which coincides with the previous one for normed spaces, and only depends on the topology: a set A is bounded if, for any sequence (x_n) of points of A, and any sequence (t_n) of
scalars tending to 0, the sequence $(t_n x_n)$ converges to 0. An
elementary argument shows that this definition is equivalent to
the following one: for any neighborhood V of 0, there is
a scalar $\lambda > 0$ such that $\lambda A \subset V$. The first general result
using this notion was the characterization of (Hausdorff) topological vector spaces for which the topology can be defined
by a norm, found by A. Kolmogoroff in 1935 [130]: they are
those for which there exists a bounded neighborhood of 0.

   Meanwhile, a new kind of topological vector spaces was introduced in 1934 by Köthe and Toeplitz [129]. For any vector
subspace E of the space $\mathbb{R}^{\mathbb{N}}$ (or $\mathbb{C}^{\mathbb{N}}$) of all real (or complex)
sequences, they consider the space E^* of all sequences (u_n)
in $\mathbb{R}^{\mathbb{N}}$ (resp. $\mathbb{C}^{\mathbb{N}}$) such that $\sum_n |u_n x_n|$ converges for all
$(x_n) \in E$ (nowadays one says that E^* is the Köthe dual of E). One
can then consider the space E^{**}, which obviously contains E;
when $E^{**} = E$, Köthe and Toeplitz say E is perfect ("vollkommen"),
and it is this kind of space which is mainly studied in their
paper, as well as in many subsequent papers of Köthe and his
pupils ([128], [130]). On such a space E, they first define
the weak topology $\sigma(E, E^*)$ in the same way as von Neumann (§1),

neighborhoods of O being defined by a finite number of ine-
qualities $|\langle x, a_j^* \rangle| \leq 1$ with $a_j^* \in E^*$ arbitrary. For that
topology, they define <u>bounded sets</u> as sets $A \subset E$ such that
each function $x \mapsto \langle x, a^* \rangle$ $(a^* \in E^*$ arbitrary) is bounded in
A, which is of course a special case of the general notion
mentioned above, although they introduce it without any reference. But their next step is particularly interesting; as E
and E^* play symmetrical parts, one can also define the weak
topology $\sigma(E^*, E)$ and bounded sets in E^*; this enables
them to define <u>on</u> E a new topology, the <u>strong topology</u>,
where neighborhoods of O are defined in the following way:
to each bounded set B in E^*, one associates the set V_B
of all $x \in E$ such that $|\langle x, y^* \rangle| \leq 1$ for all $y^* \in B$ (the
"polar set" of B in a later terminology), and the V_B
constitute a fundamental system of neighborhoods of O for
the strong topology. One can then define bounded sets in E
for that topology, and one of the chief results of Köthe and
Toeplitz is that bounded sets in E are <u>the same</u> for the weak
and strong topology.

All topological vector spaces mentioned above belonged to
what we now call <u>locally convex spaces</u>, but the general definition of these spaces (under the name "convex spaces") was
only given in 1935 by von Neumann, in view of a study of almost periodic functions [221, vol.II, p.508-527]. This co-
incided with a revival of interest in the properties of convex
sets in topological vector spaces, which after Helly had been
pretty much neglected: in Banach's book, they are only briefly mentioned in a Note at the end of the book [15, p.246].
However, in 1933, S. Mazur gave the "geometric" version of the

Hahn-Banach theorem, generalizing Minkowski's theory by showing that, if K is an open convex set in a normed space E,
there is a closed hyperplane of support of K through each
boundary point of K [158]. A little later, M. Krein and
D. Milman introduced the concept of extreme point for a convex set K, i.e. a point of K such that there is no open
line segment containing the point and contained in K; they
proved the remarkable fact that there are always "enough" extreme points for a compact convex set K, more precisely K
is the smallest closed convex set containing all the extreme
points [131]; a theorem which was to have many important applications in various domains of Functional Analysis.

Once the locally convex spaces had been defined, the Köthe-
Toeplitz procedure could be put in a more general context:
one starts with two vector spaces E, F and a bilinear form
B on ExF, which is non degenerate, i.e. such that the relation "$B(x,y) = 0$ for all $y \in F$ " is equivalent to $x = 0$
and "$B(x,y) = 0$ for all $x \in E$ " equivalent to $y = 0$. One
then considers on E all Hausdorff locally convex topologies
for which F is the dual of E; the determination of these
topologies was done by G. Mackey [154], who showed that one
gets a fundamental system of neighborhoods of 0 for such a
topology by taking the finite intersections of the "polar"
sets of a family $\mathfrak{S}$ of subsets of F, which consists of
compact symmetric sets for the weak topology $\sigma(F,E)$ and form
a covering of F; in addition, Mackey showed that for all
these topologies, the bounded sets in E are the same.

We shall not try to describe in detail the very numerous

papers devoted to topological vector spaces which have been
published since 1950. Shortly after that date appeared the
first comprehensive treatises on the subject ([26], [62], [92],
[125], [128], [215]). Most researches have been devoted to
the study of particular types of locally convex spaces, such
as Fréchet spaces and their direct limits ([55], [92]), various types of "vollkommen" sequence spaces in the sense of
Köthe-Toeplitz, which yield a rich harvest of examples and
counterexamples, as well as many types of spaces consisting
of functions with various properties. The most significant
recent results concern the various topologies which one can
define on the tensor product $E \otimes F$ of two locally convex
spaces; they were studied in depth in a remarkable paper by
A. Grothendieck, which deserves to be considered as realizing
the greatest progress in Functional Analysis after the work
of Banach [91]; this study led its author to the discovery of
a new class of locally convex spaces, the nuclear spaces,
which in a sense are much closer to finite dimensional spaces
than even Hilbert spaces (with which they have some surprising
connections) [59]. Most spaces occurring in the theory of
distributions (§3) are nuclear spaces, and nuclear Fréchet
spaces have become quite important in the theory of probability.
Finally one should mention a large literature on convex sets
in topological vector spaces, taking its origin in a beautiful
result of Choquet giving to the Krein-Milman theorem a quanti-
tative interpretation: if C is the convex hull of the union
of $\{0\}$ and a compact convex set K contained in a closed
hyperplane not containing O, then every point of C is the

barycenter of a positive measure carried by the extreme points
of C, and this measure is unique if and only if the order
relation defined by the cone C' , union of all the λC for
$\lambda > 0$, is a lattice [42]. This result has important applications in potential theory.

§3 - The theory of distributions

Between 1930 and 1940, several mathematicians began to investigate systematically the concept of "weak" solution of a
linear partial differential equation, which we have seen ap-
pearing episodically (and without a name) in Poincaré's work
(chap.III, §2). In general, let $P: f \mapsto \sum_{\alpha} a_{\alpha} D^{\alpha} f$ be any linear
differential operator with C^{∞} coefficients, defined in an open
set $\Omega \subset \mathbb{R}^n$, and write $\langle f,g \rangle = \int_{\Omega} f(x)g(x)dx$ for f locally integrable in Ω and g continuous with compact support
in Ω; then it is easy to generalize Lagrange's definition
of the __adjoint__ differential operator $^t P$, which in particular satisfies

(1) $\langle P \cdot f, g \rangle = \langle f, {}^t P \cdot g \rangle$

when f is C^{∞} in Ω, and g is C^{∞} in Ω with compact
support. If f is a C^{∞} solution of $P \cdot f = 0$, we have
therefore $\langle f, {}^t P \cdot g \rangle = 0$ for all functions g which are C^{∞}
in Ω and have compact support. Conversely, any function f
locally integrable in Ω and having that property is called
a __weak__ solution of the equation $P \cdot u = 0$, even if it is not differentiable at all, and the problem which had confronted

Poincaré was to prove that a weak solution is in fact a genuine C^∞ solution.

But in fact the same problem, for the simplest differential
operator $D = \frac{d}{dx}$, had already been considered and solved in
the affirmative by P. Du Bois-Reymond in 1879 [60]. Prodded
by Weierstrass's criticism of the Calculus of variations
(chap.II, §4), he undertook to prove that if a C^1 function
y in an interval [a,b] is an extremum for the integral
$I(y) = \int_a^b F(x,y,y')dx$, where F is a C^1 function, then
the Euler equation

$$(2) \qquad \frac{d}{dx}\left(\frac{\partial F}{\partial y'}(x,y,y')\right) - \frac{\partial F}{\partial y}(x,y,y') = 0$$

makes sense, which certainly is not obvious since nothing
guarantees a priori that $\frac{\partial F}{\partial y'}(x,y,y')$ is differentiable!
Following the classical procedure of Lagrange, one writes
that for any C^1 function ζ having compact support in]a,b[,
the function $\varepsilon \mapsto I(y+\varepsilon\zeta)$ has an extremum for $\varepsilon = 0$, which
is equivalent to the relation

$$(3) \qquad \int_a^b \left(\frac{\partial F}{\partial y}\zeta + \zeta'\frac{\partial F}{\partial y'}\right)dx = 0;$$

but instead of integrating by parts to eliminate ζ', one
instead eliminates ζ by using the fact that $\zeta(a) = \zeta(b) = 0$:
an integration by parts enables indeed to write (3) in the
form

$$(4) \qquad \int_a^b \zeta'(x)f(x)dx = 0,$$

where $f(x) = \frac{\partial F}{\partial y'}(x,y(x),y'(x)) - \int_a^x \frac{\partial F}{\partial y}(t,y(t),y'(t))dt$ is only known to be <u>continuous</u> in $]a,b[$. The problem is to prove that f is a <u>constant</u>, for then it will follow that $\frac{\partial F}{\partial y'}(x,y(x),y'(x))$ is indeed differentiable and equation (2) is satisfied. However this amounts to showing that any weak solution of $Du = 0$ is a constant, which is exactly what du Bois-Reymond proves[*].

Such a result of course could not be expected for any differential operator: for instance, if A and B are <u>any</u> locally integrable functions in $\mathbb{R}$, the function $(x,y) \mapsto A(x) + B(y)$ is a weak solution of $\frac{\partial^2 u}{\partial x \partial y} = 0$, for one has

$$\int_{\mathbb{R}} A(x)dx \int_{\mathbb{R}} \frac{\partial^2 g}{\partial x \partial y} dy = 0 \quad \text{and} \quad \int_{\mathbb{R}} B(y)dy \int_{\mathbb{R}} \frac{\partial^2 g}{\partial x \partial y} dx = 0 \quad \text{for}$$

any C^2 function g with compact support.

A step further would lead to defining a "generalized" operator P, acting on functions which were not supposed differentiable at all: for a locally integrable function f in Ω, $P \cdot f$ would be (by definition) a locally integrable function h such that, for <u>any</u> C^∞ function g in Ω with compact support, one has

$$(5) \qquad \langle h,g \rangle = \langle f, {}^t P \cdot g \rangle.$$

In a slightly different context, E. Cartan in 1922 [39] had observed that it was sometimes possible to define an "exterior derivative" $d\omega$ for a differential 2-form $\omega = Pdy \wedge dz + Qdz \wedge dx + Rdx \wedge dy$, even if P, Q, R were merely continuous but not necessarily differentiable; one would define $d\omega = Sdx \wedge dy \wedge dz$ if S was a continuous function such that.

_____

[*]It is interesting to remark that in this paper du Bois-Reymond uses (probably for the first time) what we now call "test functions", i.e. C^∞ functions with compact support.

(6) $$\iiint_V S\,dxdydz = \iint_\Sigma \left(Pdy \wedge dz + Qdz \wedge dx + Rdx \wedge dy\right)$$

for any open set V with smooth boundary Σ. As an example, he gave the form for which $P = \dfrac{\partial U}{\partial x}$, $Q = \dfrac{\partial U}{\partial y}$, $R = \dfrac{\partial U}{\partial z}$, where U is the potential of a density ρ which is only supposed to be continuous; then P, Q, R need not be differentiable, but nevertheless $S = -4\pi\rho$ satisfies (6).

The first systematic introduction of such "generalized" operators, for $P = \dfrac{\partial}{\partial x_j}$ (under the name of "quasi-dérivées") is to be found in a paper of J. Leray in 1934 [141] $^{(*)}$. In addition, Leray also introduces the process of <u>regularization</u> of a locally integrable function f by a sequence (ρ_n) of C^∞ functions with compact support tending to 0, such that $\rho_n \geq 0$ and $\int \rho_n\,dx = 1$: he shows that if f is continuous, $\rho_n * f$ is a C^∞ function which converges uniformly to f in every compact subset, and if h is continuous and is the "generalized derivative" of f, then $\rho_n * h$ is the (usual) derivative of $\rho_n * f$ $^{(**)}$.

With our present knowledge, we realize that this notion of "generalized derivative" was a natural consequence of the use of the Lebesgue integral. Progressively, analysts had become

---

$(*)$Leray's results were rediscovered independently by K. Friedrichs in 1939 [76].

$(**)$The study of integrals $\rho_n * f$ for various types of sequences of functions (ρ_n) was a favorite subject of analysts from Weierstrass to Lebesgue, under the name "singular integrals". For continuous functions ρ_n with compact support shrinking to a point, it had been systematically used by H. Weyl on Lie groups, as a substitute for the missing unit element in $L^1(G)$ ([227], vol.III, p.73).

familiar with the idea that two measurable functions which co-
incided except in a set of measure 0 were not to be distinguished from one another in most operations of Analysis. Not
so, however, for differentiation: if f is a C^{∞} function
in $\mathbb{R}$ and φ is the characteristic function of the set of
rationals, $f + \varphi$ is almost everywhere equal to f, but $f+\varphi$
has no derivative at any point, in the usual sense! Nevertheless it has of course a "generalized derivative" equal to
f', and this could throw doubts on the adequacy of the "natural" definition of a derivative in Analysis!

It is easy to see that if a function f in $\mathbb{R}$ has a "generalized derivative" h which is locally integrable, then
f is almost everywhere equal to $\int_0^x h(t)dt + c$, where c
is a constant. This shows that a continuous function may
have <u>almost everywhere</u> a derivative in the usual sense, without having a "generalized derivative", for instance an increasing function f which is not absolutely continuous,
another example of the inadequacy of the concept of derivative in the classical sense.

<u>A fortiori</u>, this also shows that a function which is dis-
continuous at a point of $\mathbb{R}$ cannot have a "generalized derivative". Nevertheless, following Dirac, theoretical physicists did not hesitate to consider that the Heaviside function
Y, equal to 0 for x < 0 and to 1 for x ≥ 0, had a
"generalized derivative", the so-called "Dirac function" δ,
which would have been equal to 0 for x ≠ 0, but such that
$\int \delta(x)dx = 1$; and they even introduced successive "derivatives" $\delta', \delta'', \ldots$ of that "function", writing "equations"
such as

(7) $$\int g(x)\delta^{(n)}(x-a)dx = g^{(n)}(a)$$

for a C^n function g, or

(8) $$\int \delta'(a-x)\delta^{(n)}(x-b)dx = \delta^{(n+1)}(a-b) \qquad [56].$$

For some time, mathematicians were puzzled by such manipulations, which eventually led to correct statements on genuine functions. The decisive step was taken in 1936 by S. Sobolev [200]: the outcome of these jugglings with non-existent "functions" was finally to define perfectly decent <u>linear forms</u> such as $f \longmapsto f^{(n)}(a)$ on the vector space $\mathcal{D}(\Omega)$ of all C^∞ functions with compact support defined in an open set $\Omega \subset \mathbb{R}^N$; Sobolev's idea was therefore to deal <u>directly</u> with such linear forms, provided one could characterize them by properties involving only genuine mathematics. As he was led to this idea by a very concrete question, the solution of Cauchy's problem for second order hyperbolic equations with general boundary conditions (see chap.IX, §5), he could see what kind of properties he needed, and give a general characterization of what he called "functionals" on $\mathcal{D}(\Omega)$, which we now call (after L. Schwartz) <u>distributions</u> on Ω: for each compact subset $K \subset \Omega$ one considers the subspace $\mathcal{D}(\Omega;K)$ of $\mathcal{D}(\Omega)$ consisting of all C^∞ functions with support in K, and this is a Fréchet space for the semi-norms

(9) $$p_{m,K}(f) = \sup_{|\alpha| \leq m, x \in K} |D^\alpha f(x)|;$$

distributions are then the linear forms T on $\mathcal{D}(\Omega)$, the restriction of which to each subspace $\mathcal{D}(\Omega;K)$ is <u>continuous</u>

for the preceding topology$^{(*)}$. Any locally integrable func-
tion F in Ω defined a distribution $f \mapsto \int_{\Omega} F(x)f(x)dx$, two
almost everywhere functions giving rise to the same distribu-
tion (which of course were the measures on Ω having a density with respect to Lebesgue measure), so that the space
$L^1_{loc}(\Omega)$ of classes of locally integrable functions was identified with a subspace of the space $\mathscr{D}'(\Omega)$ of all distributions; but more generally all <u>Radon measures</u> on Ω were particular distributions, and in particular one could define cor-
rectly the so-called "Dirac function" $x \mapsto \delta(x-a)$ as the measure ε_a: $f \mapsto f(a)$ defined by the mass $+1$ at the point a.
Sobolev pointed out that one can multiply a distribution T
by any C^∞ function g in Ω, by defining $g \cdot T$ as the distribution $f \mapsto T(gf)$; more important still, one may define
the <u>derivatives</u> $\frac{\partial T}{\partial x_i}$ of <u>any</u> distribution T as the distributions $f \mapsto -T(\frac{\partial f}{\partial x_i})$. Finally, he considered on the space
$\mathscr{D}'(\Omega)$ the weak topology $\sigma(\mathscr{D}'(\Omega), \mathscr{D}(\Omega))$, and showed that the
regularization process could also be applied to distributions:
$\rho_n * T$ is defined as the distribution $f \mapsto T(\overset{\vee}{\rho_n} * f)$, which
turns out to be the class of a C^∞ function, and $\rho_n * T$ converges weakly to T when n tends to $+\infty$; the fact that
distributions are thus limits (for the weak topology) of C^∞
functions has led some mathematicians to call them "generalized

---

$^{(*)}$Sobolev does not speak of topology, but defines convergent
sequences in $\mathscr{D}(\Omega)$ which correspond to these topologies on the
spaces $\mathscr{D}(\Omega;K)$. Another way of expressing the definition is to
consider the "direct limit" of the topologies of the spaces
$\mathscr{D}(\Omega;K)$; distributions are then the elements of the <u>dual</u> of
$\mathscr{D}(\Omega)$ when $\mathscr{D}(\Omega)$ is given that topology.

functions" [86].

During the same period, the need to "enlarge" in some way the domain of definition of operators other than differential operators was also felt in different parts of Analysis$^{(*)}$, and particularly in classical harmonic Analysis. The definition of the Fourier transform $\mathfrak{F}f$ of a function f defined in $\mathbb{R}^n$ only makes sense when $f \in L^1(\mathbb{R}^n)$; however, as soon as 1910, the Plancherel theorem (chap.VII, §6) showed that it is possible to define the operator $f \mapsto \mathfrak{F}f$ as an <u>isometry</u> of the Hilbert space $L^2(\mathbb{R}^n)$ onto itself, by extending it by continuity from its original domain of definition $L^1 \cap L^2$; in other words, for a function $f \in L^2$ which did not belong to L^1, the Fourier transform $\mathfrak{F}f$ could still be defined, but only by a limit process. Later, efforts were made to define similarly a Fourier transform $\mathfrak{F}f$ for functions f belonging to other spaces L^p; in his discussion of that problem, A. Weil observed in 1940 [226, p.118] that if Λ is the space of functions $f \in L^1 \cap L^\infty$ such that $\mathfrak{F}f$ also belongs to $L^1 \cap L^\infty$, then if two functions Φ, φ are such that

$$(10) \qquad \int \Phi(x) \cdot \overline{\mathfrak{F}f(x)} dx = \int \varphi(x) \cdot \overline{f(x)} dx$$

for <u>all</u> functions $f \in \Lambda$, it is legitimate to consider that Φ is the Fourier transform of φ.

$^{(*)}$For instance, in the Calculus of variations, one may consider that a smooth p-dimensional variety V in an $\mathbb{R}^n$ defines a linear "functional" $\psi \mapsto \int_V \psi$ in the vector space of differential p-forms on $\mathbb{R}^n$. This leads to the idea of "generalized varieties" [231] and of "currents" [49].

Much earlier, in 1911, H. Weyl, in relation with his work
on second order linear differential equations (chap.VII, §6),
had observed that if f is such that $f(x)/(1+|x|)$ is inte-
grable in $\mathbb{R}$ (for instance, if $xf(x)$ is bounded) and satisfies an additional regularity condition, then it is possible to write for f the Fourier inversion formula, provided one replaces $\mathfrak{F}f$ by a Stieltjes measure ([227],vol.I,
p. 359-360); this amounts to defining the Fourier transform
of a bounded Stieltjes measure, a definition which was explicitly given by P. Daniell in 1920 [46], and which became later a favorite tool of probabilists. Weyl's idea was developed by Hahn [96] and N. Wiener [229], and then extended by
S. Bochner to functions such that $f(x)/(1+|x|^k)$ is integrable,
where k is an <u>arbitrary</u> integer [24]. Using the fact that
by Fourier transformation derivation becomes multiplication
by x (up to a constant), Bochner proceeds as Riemann had
done for trigonometric series [182, p.245]: in order to
obtain an integrable function, he substracts from $e^{-ix\xi}$ a
function $L_k(x\xi)$ equal to the first k terms $\sum_{m=0}^{k-1} \frac{1}{m!} (-ix\xi)^m$
of the power series expansion of $e^{-ix\xi}$ in a compact neigh-
borhood of 0, and to 0 outside, and writes

$$(11) \qquad E(\xi,k) = \frac{1}{2\pi} \int_{-\infty}^{+\infty} f(x) \frac{e^{-ix\xi} - L_k(x\xi)}{(-ix)^k} dx .$$

This is of course only defined up to a <u>polynomial</u> in ξ of
degree ≤k-1; Bochner's idea would be to take a "derivative"
in some sense of $E(\xi,k)$ as "Fourier transform" of f, and
indeed he writes "symbolically"

$$f(x) \sim \int e^{ix\xi} d^k E(\xi,k)$$

what would be the "inversion formula"; he gives a definition
of such a "k-th derivative" by repeated integration by parts;
but most of the time he only works with the <u>functions</u> $E(\xi,k)$
in the applications he gives to difference equations.

It was one of the main contributions of L. Schwartz that he
saw, in 1945 [194], that the concept of distribution introduced
by Sobolev (which he had rediscovered independently) could
give a satisfactory generalization of the Fourier transform
including all the preceding ones. Instead of considering the
space Λ introduced by A. Weil, which is not easy to describe
explicitly, he had the idea to take as "test functions" the
C^∞ functions f in $\mathbb{R}^n$ which are such that f and <u>all</u> its
derivatives are "rapidly decreasing at infinity", i.e. such
that their product with any polynomial is integrable. The
essential property of that space $\mathcal{S}(\mathbb{R}^n)$ of "declining functions" is that the Fourier operator $f \mapsto \mathcal{F}f$ is a bicontinuous
<u>bijection</u> of $\mathcal{S}(\mathbb{R}^n)$ with the Fréchet topology defined by the semi-
norms

$$(12) \qquad q_{s,m}(f) = \sup_{|\alpha| \le s, \, x \in \mathbb{R}^n} (1+|x|)^m \, |D^\alpha f(x)|.$$

It is easy to see that for each compact subset K of $\mathbb{R}^n$,
the space $\mathcal{D}(\mathbb{R}^n;K)$ is contained in $\mathcal{S}(\mathbb{R}^n)$ and the injection
$\mathcal{D}(\mathbb{R}^n;K) \to \mathcal{S}(\mathbb{R}^n)$ is continuous; furthermore the union $\mathcal{D}(\mathbb{R}^n)$
of all the $\mathcal{D}(\mathbb{R}^n;K)$ is dense in $\mathcal{S}(\mathbb{R}^n)$. The continuous linear forms on $\mathcal{S}(\mathbb{R}^n)$ can thus be considered as special distributions, which Schwartz calls <u>tempered distributions</u>.
The Fourier transform $T \mapsto \mathcal{F}T$ in the space $\mathcal{S}'(\mathbb{R}^n)$ of tempered distributions (dual of $\mathcal{S}(\mathbb{R}^n)$) is then defined (by a
generalization of equation (10)) as the <u>transposed</u> automor-

phism of the Fourier transform in $\mathcal{S}(\mathbb{R}^n)$, in other words the Fourier transform $\mathcal{F}T$ of a tempered distribution T is defined by the relation

$$(13) \qquad \langle \mathcal{F}T, f \rangle = \langle T, \mathcal{F}f \rangle \quad \text{for all} \quad f \in \mathcal{S}(\mathbb{R}^n).$$

To the credit of L. Schwartz must be added his persistent efforts to weld all the previous ideas into a unified and complete theory, which he enriched by many definitions and results (such as those concerning the tensor product and the convolution of distributions) in his now classical treatise [194]. By his own research and those of his numerous students, he began to explore the potentialities of distributions, and gradually succeeded in convincing the world of analysts that this new concept should become central in all linear problems of Analysis, due to the greater freedom and generality it allowed in the fundamental operations of Calculus, doing away with a great many unnecessary restrictions and pathology [(*)]. One should reserve a particular mention to what is probably the most original of his contributions, the "kernel theorem" [195]. Ever since Hilbert's and F. Riesz's work, it had been

---

[(*)] The role of Schwartz in the theory of distributions is very similar to the one played by Newton and Leibniz in the history of Calculus: contrary to popular belief, they of course did not invent it, for derivation and integration were practiced by men such as Cavalieri, Fermat and Roberval when Newton and Leibniz were mere schoolboys. But they were able to systematize the algorithms and notations of Calculus in such a way that it became the versatile and powerful tool which we know, whereas before them it could only be handled via complicated arguments and diagrams (see [28]).

realized that integral operators $f \longmapsto K \cdot f$ defined by a
"kernel function" $K(x,y)$, as $(K \cdot f)(x) = \int K(x,y)f(y)dy$,
were very far from exhausting the general concept of linear
operator, since not even the identity could be expressed in
that manner! It is therefore very remarkable that if one
replaces "kernel functions" by "kernel distributions" in that
definition, one practically obtains all linear operators which
one meets in problems of Analysis. More precisely, if $X \subset \mathbb{R}^m$
and $Y \subset \mathbb{R}^n$ are open sets, any linear mapping K of $\mathcal{D}(X)$
into $\mathcal{D}'(Y)$, which is only supposed to yield partial contin-
uous mappings $\mathcal{D}(X;L) \rightarrow \mathcal{D}'(Y)$ for any compact subset $L \subset X$
(when $\mathcal{D}(X;L)$ is given its Fréchet topology and $\mathcal{D}'(Y)$ the
weak topology), can be defined by a uniquely determined "ker-
nel distribution" $K \in \mathcal{D}'(X \times Y)$, in such a way that for any
$u \in \mathcal{D}(X)$, the distribution $K \cdot u$ satisfies

(14) $\langle K \cdot u, v \rangle = \langle K, u \otimes v \rangle$

for any function $v \in \mathcal{D}(Y)$. The great interest of this result
lies in the fact that most spaces E of functions defined in
X are such that $\mathcal{D}(X) \subset E \subset \mathcal{D}'(X)$, the injections $\mathcal{D}(X;L) \rightarrow$
$\rightarrow E$ and $E \rightarrow \mathcal{D}'(X)$ being continuous; if A (resp. B) is
such a space of functions defined in X, and $U: A \rightarrow B$ a
continuous linear mapping, the composed map $\mathcal{D}(X;L) \rightarrow A \xrightarrow{U} B \rightarrow$
$\rightarrow \mathcal{D}'(X)$ is continuous, hence is defined by a "kernel distribution". For instance, the identity map $A \rightarrow A$ is defined
by the distribution I which is a measure carried by the
diagonal Δ_X in $X \times X$ and is such that

$$\int_{X \times X} w(x,y)dI(x,y) = \int_X w(x,x)dx.$$

CHAPTER IX

APPLICATIONS OF FUNCTIONAL ANALYSIS

TO DIFFERENTIAL AND PARTIAL DIFFERENTIAL EQUATIONS

I will not try to enumerate all the applications which have
been made of Functional Analysis in the last 50 years, and
which have amply justified the creators of that discipline.
But as we have seen in the first 4 chapters how most notions
and problems of Functional Analysis had their origin in questions relative to ordinary or partial differential equations,
I think it is worthwhile to give a sketchy description of a
few of the most conspicuous progress in those questions which
have been made by an imaginative use of the new tools provided
by Functional Analysis, mostly spectral theory and the theory
of distributions.

§1 - Fixed point theorems

In the first applications which we shall mention, however,
little more is used of Banach spaces beyond their definition,
and the results primarily concern non linear equations. The
main idea is similar to the application of the contraction
principle (chap.VI, §3) to the local existence theorems for
differential equations, by writing them in the form $z = F(z)$
for z in some Banach space E of functions; if B is the
closure of an open bounded convex set in E and F is a con-

233

traction mapping B into itself, then the contraction principle says there exists a unique solution of z = F(z) in B,
what one calls a fixed point for F. But after the first years
of the XXth century, new possibilities of obtaining "fixed
point theorems" appeared with the first results of a new
branch of mathematics, Algebraic Topology, created by H. Poincaré
in 1895-1900: using the concepts of that theory, L.E.J. Brouwer
could show in 1910 that if B is homeomorphic to a closed ball
in some finite dimensional vector space E, and F is any
continuous map of B into itself (which is not supposed any
more to be a contraction), then F has at least one fixed
point in B.

The problem was to find a similar theorem applicable to in-
finite dimensional Banach spaces E. The first result in that
direction was obtained in 1922 by G.D. Birkhoff and O. Kellogg,
who considered the case in which $E = C(I)$ or $E = \ell^2$, and
showed that Brouwer's theorem could be extended, provided one
took for B a compact convex set [22]; their fundamental device consists in using the compactness of B to "approximate"
it by a finite dimensional compact convex set B_n, and to similarly "approximate" F by a continuous mapping F_n of B_n
into itself, to which Brouwer's theorem may be applied. This
method was taken up and greatly expanded by J. Schauder, who
showed that it could be applied to any Banach space E, and
also, for separable Banach spaces, that one could replace compactness of B and continuity of F by weak compactness and
weak continuity ([186], [187]). This enabled him to prove,
for instance, existence of a solution of the equation

(1)
$$\Delta z = f(x,y,z, \frac{\partial z}{\partial x}, \frac{\partial z}{\partial y})$$

in a domain $\Omega \subset \mathbb{R}^2$ with smooth boundary (no connected component of which is reduced to a point), vanishing at the boundary, under the only assumption that f is bounded and continuous for bounded values of the 5 variables on which it depends; the method consists in transforming (1) into an integro-differential equation

(2) $\quad z(x,y) = \iint_{\Omega} G(x,y,\xi,\eta)f(\xi,\eta,z(\xi,\eta), \frac{\partial z}{\partial \xi}, \frac{\partial z}{\partial \eta})d\xi \, d\eta$

where G is the Green function for Ω.

A little later, a much more sophisticated approach enabled Schauder to solve Cauchy's problem locally for quasi-linear hyperbolic equations[*]

(3)
$$\sum_{i,k} A_{ik}(x_1,\ldots,x_n,z,\frac{\partial z}{\partial x_1},\ldots,\frac{\partial z}{\partial x_n}) \frac{\partial^2 z}{\partial x_i \partial x_k} =$$
$$= A(x_1,\ldots,x_n,z,\frac{\partial z}{\partial x_1},\ldots,\frac{\partial z}{\partial x_n}).$$

His method consists, for a <u>given</u> function $z(x_1,\ldots,x_n)$, to solve the Cauchy problem for the <u>linear</u> hyperbolic equation in the unknown function Z

(4)
$$\sum_{i,k} A_{ik}(x_1,\ldots,x_n,z,\frac{\partial z}{\partial x_1},\ldots,\frac{\partial z}{\partial x_n}) \frac{\partial^2 Z}{\partial x_i \partial x_k} =$$
$$= A(x_1,\ldots,x_n,z,\frac{\partial z}{\partial x_1},\ldots,\frac{\partial z}{\partial x_n}).$$

---

[*]The equation is supposed to be "normal", which means that the left hand side is such that the quadratic form $\sum_{i,k} A_{ik}\xi_i\xi_k$ has signature $(1,n-1)$.

When   z   is in a suitable set   K,   the problem has a unique

solution $Z(z)$,   and existence of a solution   z   of (3) in K

will be obtained if one shows that the equation $Z(z) = z$ has

a solution in   K.   The problem consists in choosing   K   such

that the extension of Brouwer's fixed point theorem is appli-

cable; the main point is to obtain a priori inequalities   for

solutions of "normal" linear hyperbolic equations

$$L(u) \equiv \sum_{i,k} A_{ik}(x_1,\ldots,x_n)\frac{\partial^2 u}{\partial x_i \partial x_k} + \sum_j B_j(x_1,\ldots,x_n)\frac{\partial u}{\partial x_j}$$

(5)

$$+ C(x_1,\ldots,x_n)u = F(x_1,\ldots,x_n)$$

defined in a truncated pyramid   P   having its larger base   B

in $\mathbb{R}^{n-1}$.   Using a method first introduced in 1926 by

Friedrichs and H. Lewy, which consists in transforming the

integral $\int_P \frac{\partial u}{\partial x_n} L(u)d\omega$ by integration by parts and Stokes'

formula, Schauder obtains (for suitable restrictions on   P

and the coefficients A_{ik}, B_j and   C   in (5)) an inequality

$$\int_P ((\frac{\partial u}{\partial x_1})^2 +\ldots+ (\frac{\partial u}{\partial x_n})^2)d\omega \leq$$

(6)

$$\leq M(\int_B (u^2 + (\frac{\partial u}{\partial x_1})^2 +\ldots+ (\frac{\partial u}{\partial x_n})^2)d\sigma + \int_P F^2 d\omega)$$

with a constant   M   independent of   u;   by differentiation he

gets similar inequalities for all derivatives (of any order)

of   u.   The space   E   is then defined by the norm

$\sup_{x \in P} (\sum_{|\alpha| \leq r} |D^\alpha f(x)|)$ for a suitable value of   r,   but the

set   K   is a ball in   E   for another norm, namely

$(\int_P (\sum_{|\alpha| \leq s} |D^\alpha f(x)|^2)dx)^{1/2}$ for another value of   s;   his

a priori inequalities enable then Schauder to show that K
is compact in E $^{(*)}$ and $z \mapsto Z(z)$ continuous in K [190].

Another of the famous theorems proved by Brouwer was the
invariance of domain: if Ω is an open subset of $\mathbb{R}^n$, and
F an injective continuous map of Ω into $\mathbb{R}^n$, then the
image $F(\Omega)$ is again an open subset. In 1929, Schauder
showed that the theorem was still true in some types of Banach
spaces, for maps F of the type $x \mapsto x+H(x)$, where H is
completely continuous in the sense of F. Riesz, but not necessarily linear. From this he deduced for instance that if one
knows that the equation

(7) $\Delta z - f\left(x, y, z, \frac{\partial z}{\partial x}, \frac{\partial z}{\partial y}\right) = \psi(x, y)$

has at most a solution taking given values $\varphi(s)$ at the
boundary of $\Omega \subset \mathbb{R}^2$, then, if for given functions φ_o, ψ_o
there exists such a solution, the same is true for functions
φ, ψ sufficiently close to φ_o, ψ_o [189].

But the most sophisticated application of Algebraic Topology
to functional equations was made in the famous 1934 paper by
J. Leray and J. Schauder [142]. If U is an open set in $\mathbb{R}^n$,
such that $\bar{U}$ is compact, and f is a continuous mapping of
$\bar{U}$ in $\mathbb{R}^n$, then, for each point $z \in \mathbb{R}^n$ which does not belong to the image by f of the boundary of U, Brouwer had
shown that one may attach an integer $d(f, U, z)$ which only
depends on the connected component of $\mathbb{R}^n - f(Fr(U))$ to which
z belongs, and varies continuously with λ when $f(x) = F(x, \lambda)$

_____

$^{(*)}$This is probably the first time the norm of the space H^s
   makes its appearance.

where F is continuous and λ a real parameter; furthermore,
when $d(f,U,z) \neq 0$, the inverse image $f^{-1}(z)$ is not empty.
Leray and Schauder were able, by approximating compact subsets of a Banach space by finite dimensional subsets, to prove,
by application of this theorem of Brouwer, the following existence theorem. Let E be a Banach space, $I \subset R$ a compact
inverval, $\Omega \subset E \times I$ a bounded open set in $E \times I$, $F: \bar{\Omega} \to E$ a
mapping which is completely continuous, and in addition uni-
formly continuous. One assumes that for every $\lambda \in I$, there
are no solution of the equation $x - F(x,\lambda) = 0$ in the boundary
of Ω , and that, for <u>one</u> value $\lambda_o \in I$, the equation
 $x - F(x,\lambda_o) = 0$ has exactly one solution in Ω ; then, for
<u>every</u> $\lambda \in I$, there exists <u>at least one</u> solution of $x - F(x,\lambda) =$
 $= 0$ in Ω .

The application of that theorem to partial differential
equations usually necessitates subtle <u>a priori</u> inequalities
which guarantee that all the assumptions of the theorem are
satisfied.

§2 - Carleman operators and generalized eigenvectors

For applications to partial differential equations, it is
necessary to generalize the notion of Carleman operators defined in chap.V, §3. Let X be a locally compact metrizable
and separable space, μ a positive measure on X and H a
separable Hilbert space (finite dimensional or not); let
 $(a_n)_{n \in J}$ be a Hilbert basis of H , where J is a finite or
denumerable set. One defines a new Hilbert space $L^2_H(X,\mu)$ as
the space of <u>vector valued</u> functions $f: x \longmapsto \sum_{n \in J} f_n(x)a_n$,

mappings of X into H such that each f_n is a function of

$L^2(X,\mu)$ with complex values, and that $|f|^2 = \sum\limits_{n \in J} |f_n|^2$ is

μ-integrable; the scalar product in that Hilbert space is

then defined by

$$(8) \qquad\qquad (f|g) = \sum_{n \in J} \int_X f_n \bar{g}_n d\mu \quad .$$

Let now Y be another locally compact metrizable and sepa-

rable space, ν a positive measure on Y. A <u>Carleman kernel</u>

on X×Y (for the measure $\mu \otimes \nu$ and the Hilbert space H)

is then a mapping K: $(x,y) \mapsto (K_n(x,y))_{n \in J}$ of X×Y into $\mathbb{C}^J$

such that:

  1º   each complex function K_n is $(\mu \otimes \nu)$-measurable;

  2º   there is a null set $N \subset Y$ such that, for each $y \notin N$,

the function $x \mapsto K_n(x,y)$ is μ-measurable and the function

$x \mapsto \sum\limits_{n \in J} |K_n(x,y)|^2$ is μ-integrable.

  If $f = \sum\limits_{n \in J} f_n a_n$ is a function of $L_H^2(X,\mu)$, the function

$x \mapsto \sum\limits_{n \in J} K_n(x,y)f_n(x)$ is μ-integrable for all $y \notin N$, and

the function

$$(9) \qquad\qquad g(y) = \int_X \sum_{n \in J} K_n(x,y)f_n(x)d\mu(x)$$

defined in Y-N, is ν-measurable. One writes $K \cdot f = g$, and

K, defined in $L_H^2(X,\mu)$ is called the <u>Carleman operator</u> de-

fined by the Carleman kernel K.

  In 1952, F. Mautner discovered that Carleman operators can

be <u>characterized</u> by properties which are independent of the

definition by a kernel [157]: suppose that there is a null

set $N \subset Y$ and, for each $y \in Y-N$, a <u>continuous linear map</u>

F_y: $L^2_H(X,\mu) \to \mathbb{C}$ such that for any function $f \in L^2_H(X,\mu)$, the
map $y \mapsto F_y(f)$ is ν-measurable.   Then there is a Carleman
kernel $K = (K_n)$ and a null set $N' \supset N$ such that, for any
$y \notin N'$, $F_y(f) = (K \cdot f)(y)$ for all functions $f \in L^2_H(X,\mu)$.

We have seen (chap.VII, §5) that a continuous normal operator has a "diagonalization" which transforms it into multiplication   by   the function "identity" $\zeta \mapsto \zeta$.   More generally, one defines a diagonalization of a continuous normal
operator B in a Hilbert space E for a function $\zeta \longmapsto \Phi(\zeta)$
as in chap.VII, §5, by replacing $\mathrm{Sp}(B)$ by a separable locally compact space   Y,   the function $\zeta \mapsto \zeta$ being replaced
by a mapping $\zeta \mapsto \Phi(\zeta)$ of   Y   into $\mathbb{C}$.

With the help of his characterization of Carleman operators,
Mautner was able to get a much more precise description of
such a diagonalization when the operator B is defined in a
Hilbert space $E = L^2(X,\mu)$,   and is a Carleman operator corresponding to a Carleman kernel $(x,y) \mapsto K(x,y)$ defined in
$X \times X$,   relative to the measure $\mu \otimes \mu$;   in addition one assumes
that for the isometry $T = (T_j)_{1 \le j < \omega}$ defining the diagonalization for the function Φ,   one has $\Phi(\zeta) \ne 0$ almost everywhere in   Y   (for the measure ν),   which is equivalent to
assuming that B and B^* are injective.   It is then possible
to describe the isometries

$$T_j \colon E_j \to F_j = L^2(Y, \varphi_{S_j} \cdot \nu)$$

performing that diagonalization, in the following way:

1º   For each index   j   such that $1 \le j < \omega$,   there is a
$(\mu \otimes \nu)$-measurable function $(x,\zeta) \mapsto e_j(x,\zeta)$ such that

$e_j(x,\zeta) = 0$ for $\zeta \notin S_j$, and that for almost all $x \in X$, the function

$$\zeta \longmapsto |\Phi(\zeta)|^2 \sum_{1 \le j < \omega} |e_j(x,\zeta)|^2$$

is ν-integrable.

2º Let $r(x) = (\int_X |K(x,y)|^2 \, d\mu(y))^{1/2}$, which is μ-measurable and almost everywhere finite; then, for every function $f \in E_j$ such that $\int_X^* r(x)|f(x)| \, d\mu(x) < +\infty$ one has

(10) $$(T_j \cdot f)(\zeta) = \int_X f(x)\overline{e_j(x,\zeta)} \, d\mu(x)$$

for almost every $\zeta \in Y$.

3º There is a null set $N \subset X$ having the following property: for every j such that $1 \le j < \omega$, let u_j be a function belonging to F_j, and suppose that the function

$$\zeta \longmapsto |\Phi(\zeta)|^{-2} \sum_{1 \le j < \omega} |u_j(\zeta)|^2$$

is ν-integrable. Then, for every $x \notin N$, the function $\zeta \longmapsto \sum_{1 \le j < \omega} e_j(x,\zeta)u_j(\zeta)$ is ν-integrable, and

(11) $$(T^{-1} \cdot (u_j))(x) = \int_Y (\sum_{1 \le j < \omega} e_j(x,\zeta)u_j(\zeta)) \, d\nu(\zeta).$$

The nature of this result is better understood when one specializes it to a situation stemming from harmonic Analysis. Take $X = G$, a separable commutative locally compact group; let μ be a Haar measure on G, and consider a complex function $b \in L^1(G) \cap L^2(G)$; then $B: f \longmapsto b*f$ is a continuous <u>normal</u> operator in $L^2(G)$, such that $B^* \cdot f = \overset{\vee}{b}*f$; furthermore, from the definition

$$(b*f)(x) = \int_G b(x-y)f(y)d\mu(y)$$

it follows that B is a <u>Carleman operator</u> corresponding to
the Carleman kernel $K(x,y) = b(x-y)$. The Plancherel theorem
and the multiplicative property of the Fourier transform
(chap.VII, formula (44)) show that the isometry $T: f \longmapsto \mathcal{F}f$
defines a <u>diagonalization</u> of the operator B, with $Y = \hat{G}$,
$\omega = 2$, $\Phi(\zeta) = \mathcal{F}b(\zeta)$, and ν a Haar measure on $\hat{G}$. Formulas
(10) and (11) then boil down to

$$(12) \quad (T \cdot f)(\zeta) = \int_G f(x)\overline{e(x,\zeta)}d\mu(x), \qquad (T^{-1} \cdot u)(x) = \int_{\hat{G}} e(x,\zeta)u(\zeta)d\nu(\zeta)$$

with $e(x,\zeta) = \langle x,\zeta \rangle$ for $x \in G$ and $\zeta \in \hat{G}$, i.e. the definitions of the <u>Fourier transform</u> and of its inverse; these
formulas are only valid for $f \in L^1(G) \cap L^2(G)$ and
$u \in L^1(\hat{G}) \cap L^2(\hat{G})$, which shows that the restrictions imposed
on the functions f and u_j in (10) and (11) cannot be completely suppressed. Finally, for every $\zeta \in \hat{G}$, one has

$$(13) \qquad\qquad b*e(\cdot,\zeta) = \Phi(\zeta)e(\cdot,\zeta)$$

and although the functions $e(\cdot,\zeta)$ <u>do not</u> belong to $L^2(G)$
in general, they are in some sense "generalized eigenvectors"
for the operator B, exhibiting the same phenomenon already
observed by F. Riesz (chap.VII, §2).

In 1953, it was simultaneously observed by L. Gårding [81]
and F. Browder [34] that this phenomenon of "generalized eigenvectors" occurs for all self-adjoint operators stemming
from <u>formally self-adjoint elliptic differential operators</u>
(see §5). It is assumed that such an operator P or order m
in $L^2(X)$ (where X is an open bounded subset of $\mathbb{R}^n$) pos-

sesses a __self-adjoint__ extension A_P; then $L = A_P + iI$ (I identity) is a __normal__ unbounded operator, which is a bijection of $\mathrm{dom}(A_P)$ onto $L^2(X)$; the inverse L^{-1} is therefore a __continuous__ normal operator in $L^2(X)$, and the same is true of course of its iterates $B = L^{-q}$. It follows from the existence of a parametrix of P (see §5) that for $qm > n$, B is a __Carleman operator__; it then also follows from the hypoellipticity of P (see §5) and from Mautner's theorem that there is a diagonalization of A_P with $Y = \mathrm{Sp}(A_P)$ and $\Phi(\zeta) = \zeta$, for which (with the preceding notations) each function $e_j(\cdot,\zeta)$, for $\zeta \notin \mathrm{Sp}(A_P)$, is a C^∞ function (generally __not__ in $L^2(X)$) solution of the partial differential equation

$$(14) \qquad (P \cdot e_j(\cdot,\zeta))(x) = \zeta\, e_j(x,\zeta).$$

§3 - Boundary problems for ordinary differential equations

The results of H. Weyl on the spectral theory of second order linear differential equations (chap.VII, §3) naturally raised the question of their generalization to linear differential equations of arbitrary order, but that problem was only attacked by K. Kodaira in 1949 [126]. Surprizingly enough, although Stone had shown in his book [207] how von Neumann's spectral theory could be applied to yield H. Weyl's results, Kodaira elected to follow Weyl's method, suitably extended. Simultaneous work by Glazman and Neumark, and later papers by many authors completed Kodaira's results and also inserted them within von Neumann's theory; we refer the reader to

[62, vol.II, p.1588-1592] for more historical details, and we
will only describe the main features of the theory.

One considers a differential operator of even order 2r

(15) $L : u \mapsto D^r(p_0 D^r u) + D^{r-1}(p_1 D^{r-1} u) + \ldots + D(p_{r-1} Du) + p_r u$

where $p_0, p_1, \ldots, p_r$ are <u>real</u> C^∞ functions in an open interval $J = {]}\alpha, \beta{[}$ (bounded or not) of R, and $p_0(t) \neq 0$ for
all $t \in J$; it is <u>formally self-adjoint</u>, i.e. for any two
functions u, v of $\mathcal{D}(J)$ (space of C^∞ functions in J
with compact support)

(16) $(L \cdot u | v) = (u | L \cdot v)$

the scalar product being taken in $L^2(J)$. We write T_L the
operator L , considered as a hermitian (unbounded) operator
in $L^2(J)$ with $\mathrm{dom}(T_L) = \mathcal{D}(J)$. Then the adjoint T_L^* is
densely defined; more precisely, $\mathrm{dom}(T_L^*) = H_L$ is the space
of all functions u of class C^{2r-1} in J such that the
<u>distribution</u> $L \cdot u$ is a <u>function</u> of $L^2(J)$, with $T_L^* \cdot u =$
$= L \cdot u$. The von Neumann spectral theory (chap.VII, §4) shows
that H_L is the direct sum of $\mathrm{dom}(T_L^{**})$, E_L^+ and E_L^-,
where $E_L^\pm$ is the subspace of functions u of class C^{2r-1}
which are solutions of $L \cdot u = \pm\, iu$ and in addition are
<u>square integrable</u> in J; in fact, they are of class C^∞. Due
to the fact that the p_j are real functions, both spaces
E_L^+ and E_L^- have the <u>same</u> dimension $p \leq 2r$. The <u>dual</u> $\mathfrak{M}$
of $E_L^+ \oplus E_L^-$ can be identified with the space of linear forms
on H_L which vanish on $\mathrm{dom}(T_L^{**})$; it is the direct sum
$\mathfrak{M}_\alpha \oplus \mathfrak{M}_\beta$, where $\mathfrak{M}_\alpha$ (resp. $\mathfrak{M}_\beta$) is the subspace of $\mathfrak{M}$ consisting of all forms θ such that $\theta(u) = 0$ for all func-

tions $u \in H_L$ which vanish in a neighborhood of α (resp. β). If one writes the Lagrange "adjunction" formula (chap.I, §1, formula (5))

$$v(L \cdot u) - u(L \cdot v) = \frac{d}{dt}(C(u,v))$$

for functions u, v which are C^∞ in J, then it can be shown that any linear form $\theta \in \mathfrak{M}_\alpha$ (resp. $\mathfrak{M}_\beta$) can be written

$$\theta(u) = \lim_{t \to \alpha} C(u,w)(t) \qquad (\text{resp.} \quad \theta(u) = \lim_{t \to \beta} C(u,w)(t))$$

for a C^∞ function w in J for which the limit exists for every $u \in H_L$, and conversely any such function defines a linear form in $\mathfrak{M}_\alpha$ (resp. $\mathfrak{M}_\beta$). Due to this result, one says that for any $\theta \in \mathfrak{M}$, the relation $\theta(u) = 0$ is a <u>boundary condition</u> for the differential operator L.

As the defects of T_L^{**} are both equal to p, there exist <u>self-adjoint extensions</u> A_L of T_L (infinitely many if $p > 0$); for each of them, $\text{dom}(A_L)$ is a subspace of H_L of codimension p, defined by p independent "limit conditions" $\theta_j(u) = 0$ with $\theta_j \in \mathfrak{M}$ (the linear forms θ_j are not arbitrary, since one must have $(T_L^* \cdot u | v) = (u | T_L^* \cdot v)$ for u and v in $\text{dom}(A_L)$). For a given self-adjoint extension A_L, let S be its spectrum, which is a closed subset of $\mathbb{R}$ (it may be the whole line $\mathbb{R}$, and it is always infinite and unbounded). For any $\zeta \notin S$, $(A_L - \zeta I)^{-1}$ is a continuous normal operator in $L^2(J)$; one shows that it is a <u>Carleman operator</u>, with kernel $(s,t) \mapsto G(\zeta,s,t)$, called the <u>Green function</u> of $A_L - \zeta I$; one has

$$(16) \qquad\qquad G(\zeta,t,s) = \overline{G(\bar\zeta,s,t)}.$$

For each $\zeta \notin S$ and $t \in J$, the function $s \mapsto G(\zeta,t,s)$ is a C^{2r-2} function, and in each interval $]\alpha,t[$, $]t,\beta[$, it is a C^{∞} solution of the equation $L \cdot u - \zeta u = 0$, satisfying the boundary conditions which define A_L, and in addition the function $s \mapsto \dfrac{\partial^{2r-1}}{\partial s^{2r-1}} G(\zeta,t,s)$ has limits on the left and on the right at the point $s = t$, such that

$$(17) \qquad \frac{\partial^{2r-1}}{\partial s^{2r-1}} G(\zeta,t,t-) - \frac{\partial^{2r-1}}{\partial s^{2r-1}} G(\zeta,t,t+) = 1/p_o(t).$$

These conditions obviously generalize those seen in chap.VII, §3 for second order equations; they completely determine G once a fundamental system of $2r$ solutions $t \mapsto v_j(\zeta,t)$ of $L \cdot u - \zeta u = 0$ is known, and it is easily seen that one can write in matrix notation

$$(18) \quad \begin{cases} G(\zeta,s,t) = \vec{v}(\bar{\zeta},s)^* \, W^-(\zeta)\vec{v}(\zeta,t) & \text{for} \quad t < s \\[2mm] G(\zeta,s,t) = \vec{v}(\bar{\zeta},s)^* \, W^+(\zeta)\vec{v}(\zeta,t) & \text{for} \quad t > s \end{cases}$$

where $\vec{v}(\zeta,t)$ is the one column matrix $(v_j(\zeta,t))_{1 \le j \le 2r}$, $W^-(\zeta)$ and $W^+(\zeta)$ are two square matrices of order $2r$ which only depend on ζ. Furthermore, if the v_j have been chosen so that for each $t \in J$, the functions $\zeta \mapsto v_j(\zeta,t)$ are <u>holomorphic</u> in an open subset $H \subset \mathbb{C}$ $(1 \le j \le 2r)$, then W^- and W^+ are <u>holomorphic in</u> $H \cap (\mathbb{C}-S)$.

As the operator L is elliptic and formally self-adjoint, one can apply to it the Gårding-Browder theorem (§2). It can be shown that, with the notations introduced in §2, one has $\omega \le 2r+1$, in other words, A_L has <u>at most multiplicity</u> $2r$ in its spectrum S; for convenience, if $\omega < 2r+1$, one defines the function $e_j(t,\xi)$ to be identically 0 for

$1 \le j \le 2r$, $t \in J$ and $\xi \in S$; for the other values of j, $(t,\xi) \longmapsto e_j(t,\xi)$ is a $(\lambda \otimes \nu)$-measurable function in $J \times S$, such that for each $\xi \in S$, $t \longmapsto e_j(t,\xi)$ is a solution of the equation $L \cdot u - \xi u = 0$ and for almost all $t \in J$, the function $\xi \longmapsto e_j(t,\xi)$ is square integrable (for ν) in each compact subset of S; in addition, one has $e_j(t,\xi) = 0$ for $\xi \notin S_j$. For $\xi \in S \subset \mathbb{R}$, one may write (with $\vec{e}(t,\xi) =$ $= (e_j(t,\xi))_{1 \le j \le 2r}$, one column matrix)

$$(19) \qquad\qquad \vec{e}(t,\xi) = Q(\xi) \cdot \vec{v}(\xi,t)$$

and the elements of the matrix Q are ν-measurable and square integrable in every compact subset of S. Let

$$(20) \qquad\qquad P(\xi) = Q(\xi)^* \, Q(\xi)$$

which is a positive hermitian matrix for all $\xi \in S$. The spectral decomposition of the operator $(A_L - \zeta I)^{-1}$ can then be written __explicitly__ in the following way: for any function $f \in \mathfrak{D}(J)$, write

$$(21) \qquad (U \cdot f)(\zeta) = \Big(\int_J \overline{f(t)} v_j(\zeta,t) dt \Big)_{1 \le j \le 2r} \quad \begin{array}{l}\text{(one column} \\ \text{matrix);}\end{array}$$

then, for f, g in $\mathfrak{D}(J)$, one has

$$(22) \qquad (f \mid g) = \int_S ((U \cdot f)(\xi))^* \, P(\xi)((U \cdot g)(\xi)) d\nu(\xi)$$

and

$$(23) \qquad ((A_L - \zeta I)^{-1} \cdot f \mid g) = \int_S (\xi - \zeta)^{-1} ((U \cdot f)(\xi))^* P(\xi)((U \cdot g)(\xi)) d\nu(\xi).$$

The set $S_j - S_{j+1}$ is then the set of $\xi \in S$ such that __the matrix__ $P(\xi)$ __has rank equal to__ j, and the (vector) measure $P \cdot \nu$ can be recovered from the knowledge of the matrix W^+

(formula (18)) by the relation

$$
(24) \quad
\begin{aligned}
&\int_{[a,b]} P(\xi)\,d\nu(\xi) \ - \ \frac{1}{2}\,(P(a)\nu(\{a\}) + P(b)\nu(\{b\})) = \\
&\qquad = \frac{1}{2i\pi}\,\lim_{\varepsilon \to 0}\int_a^b (W^+(\sigma+i\varepsilon) - W^+(\sigma-i\varepsilon))\,d\sigma .
\end{aligned}
$$

These results had previously been obtained for second order operators by Titchmarsh [212].

Much work has been done to determine the spectrum S, the various subsets S_j, and the measure $P \cdot \nu$ under various hypotheses on the operator L ([62], [166]). It should how-ever be stressed that the behavior of the measure $P \cdot \nu$ on $\mathbb{R}$ is essentially <u>arbitrary</u>: in a remarkable paper, Gelfand and Levitan have shown in 1951 [84] that, given on an arbitrary compact subset H of $\mathbb{R}$ an arbitrary measure ρ, it is always possible to find a second order operator L <u>with</u> C^∞ <u>coefficients</u> such that $P(\xi)$ has the form $\begin{pmatrix} p_{11}(\xi) & 0 \\ 0 & 0 \end{pmatrix}$ and the restriction of the measure $p_{11} \cdot \nu$ to H is the given measure ρ.

§4 − Sobolev spaces and <u>a priori</u> inequalities

Until 1940, there was no general theory of linear partial differential equations (or systems of such equations) of <u>arbi-trary order</u>. With the exception of a few special types of equations with constant coefficients (such as the "biharmo-nic" equation $\Delta^2 u = 0$), the bulk of papers were concerned with <u>second order</u> equations in any number of variables, to which must be added a much smaller number of results on equations of arbitrary order in <u>two</u> independent variables.

When mathematicians began to be interested in "weak" solutions, and later with the arrival of the theory of distributions (chap. VIII), the scope of the theory of linear partial differential equations was greatly widened; if

$$(25) \qquad\qquad u \longmapsto P \cdot u = \sum_{|\alpha| \leq m} a_\alpha D^\alpha u$$

is a linear differential operator with complex C^∞ coefficients[*] in an open subset Ω of $\mathbb{R}^n$, then for any distribution $T \in \mathcal{D}'(\Omega)$, each product $a_\alpha D^\alpha T$ is defined, hence also $P \cdot T$, and it makes sense to ask for solutions T of the equation

$$(26) \qquad\qquad P \cdot T = S$$

where S is any given distribution in $\mathcal{D}'(\Omega)$. In particular, one may take for S a C^∞ function in Ω, and then one asks if this imposes conditions on the distributions T solutions of (26). In some cases, solutions of $P \cdot T = 0$ may be distributions of arbitrary order (i.e. as "irregular" as possible); this happens for instance for $P = \dfrac{\partial^2}{\partial x \partial y}$, where not only arbitrary locally integrable functions $A(x) + B(y)$ are solutions, but also arbitrary distributions of type $A \otimes 1 + 1 \otimes B$, where A and B are arbitrary distributions of $\mathcal{D}'(\mathbb{R})$. On the other hand, it may happen that for <u>all</u> C^∞ functions f

---

[*]The interest shown to equations with C^∞ coefficients (or C^r coefficients generally) is chiefly due to the pioneering efforts of Hadamard, who repeatedly emphasized that for applications to Physics it was unreasonable to study exclusively equation with analytic coefficients [93].

in Ω, <u>all</u> solutions of $P \cdot T = f$ are necessarily C^∞ functions; such operators P are now called <u>hypoelliptic</u>. This
is the case, for instance, when $n = 1$ and $P = D^p + a_1 D^{p-1} +$
$+ \ldots + a_p$ is any linear differential operator with leading
coefficient 1, an elementary result which follows by induction
from the case $p = 1$, which is du Bois-Reymond's lemma (chap.
VIII, §3). In 1927, S. Zaremba [233] proved a result which,
in modern language, means that the laplacian Δ is hypoelliptic, and H. Weyl in 1940 gave another proof of that
result ([227], vol.III, p.758-791).

After 1950, such questions, as well as extensions of the
classical boundary problems, began to be studied for operators
(25) of <u>arbitrary</u> order, heralding a period of unprecedented
expansion in the theory of partial differential equations.
Among the many methods developed during that period, we shall
postpone to §5 those linked to the concepts of elementary solution and parametrix, and consider here the applications of
the "<u>a priori</u> inequalities" which were made possible by the
appearance of new tools linked to the theory of distributions,
the <u>Sobolev spaces</u> and their generalizations.

We have already seen (§1) that Schauder had considered the
space of functions f of class C^p in Ω such that all de-
rivatives $D^\alpha f$ for $|\alpha| \leq p$ belong to $L^2(\Omega)$, and had used
the norm $\|f\|_p = (\int_\Omega (\sum_{|\alpha| \leq p} |D^\alpha f(x)|^2) dx)^{1/2}$ on that space,
which unfortunately was not complete for that norm. In 1936,
Sobolev had the idea of considering the functions $f \in L^2(\Omega)$
which have <u>weak</u> (= distributional) derivatives $D^\alpha f$ belong-
ing also to $L^2(\Omega)$ for $|\alpha| \leq p$, and this time this space

(with the same norm) $H^p(\Omega)$ is complete (i.e. a Hilbert space); moreover, Sobolev observed that the larger the number p, the more regular are the functions of $H^p(\Omega)$; for $p > [\frac{n}{2}] + 1$, they are functions of class C^r with $r = p - [\frac{n}{2}] - 1$, hence the intersection of all $H^p(\Omega)$ consists of the functions of class C^∞ such that all their derivatives are in $L^2(\Omega)$ [201].

Later, it was realized that the $H^s(\mathbb{R}^n)$ could be defined using the Fourier transform, as the space of distributions $T \in S'(\mathbb{R}^n)$ such that the Fourier transform $\mathcal{F}T$ is a locally integrable function for which the function $\xi \longmapsto (1 + |\xi|^2)^s |\mathcal{F}T(\xi)|^2$ is integrable. It is then clear that the definition may be extended to <u>all real numbers</u> s.

These properties were the source of what one may call the "bootstrap" method to prove that a function is C^∞: it is enough to show that <u>if</u> it belongs to <u>some</u> Sobolev space $H^r(\Omega)$, it <u>also</u> belongs to $H^{r+1}(\Omega)$. The first idea of this method is apparently due to K. Friedrichs [77], who applied it to prove that elliptic operators (25) (see §5) are hypoelliptic, a question which we postpone until §5; his tool is a new type of <u>a priori</u> inequality, which was the starting point of a very large number of similar results, for elliptic[*] and other types of operators. We shall only mention here one of the most refined ones [117, p.207]; it concerns what are called "principally normal" operators P, which we shall not attempt to describe more precisely here, but which include operators with real coefficients and operators with constant

_____

[*]For a description of the various methods based on <u>a priori</u> inequalities in 1956, see [168].

coefficients; under assumptions on Ω in relation with the characteristic hyperplanes of the operator P, too complicated to reproduce here, the fundamental result is that if u is a distribution on Ω having a compact support K, and such that $P \cdot u$ belongs to some $H^s(\Omega)$, then u necessarily belongs to $H^{s+m-1}(\Omega)$, and there is a constant $C_{s,K}$ independent of u and such that

$$(27) \qquad \| u \|_{s+m-1} \leq C_{s,K}(\| P \cdot u \|_s + \| u \|_{s+m-2}).$$

The "bootstrap" method shows that if $P \cdot u = f$ where f is a C^∞ function, then u itself is a C^∞ function. But from inequality (27) one may derive much more information: for instance the space of solutions of $P \cdot u = 0$ having support in K is finite dimensional and consists of C^∞ functions; if f is a C^∞ function such that $\int_\Omega f(x)u(x)dx = 0$ for all these functions u, then there is a C^∞ function v in Ω solution of the adjoint equation $^t P \cdot v = f$ in K (ibid., p.210). Similar uses of such inequalities have been successful in proving existence and uniqueness of Cauchy problems for hyperbolic (see §5) equations of arbitrary order ([62], vol.II, p.1748-1766).

§5 - Elementary solutions, parametrices and pseudo-differential operators

We have seen (chap.II, §3) that from the beginning the idea of "newtonian" potential was closely related to the laplacian operator; if one considers in $\mathbb{R}^3$ the integral operator U

defined by the kernel $-\frac{1}{4\pi|x-y|}$ (where $|x|$ is the eucli-
dean norm), then Poisson's equation (chap.II, formula (12))
can be written $\Delta\cdot(U\cdot\rho) = \rho$ for $\rho \in \mathcal{D}(\mathbb{R}^3)$, and one also may
write $U\cdot(\Delta\cdot f) = f$ if f is the potential having density ρ,
so that this equation is also valid for all $f \in \mathcal{D}(\mathbb{R}^3)$. Later,
when the existence of the Green function G was proved for a
domain Ω in $\mathbb{R}^3$, one had similarly the relations $\Delta\cdot(U\cdot f) =$
$= f$ and $U\cdot(\Delta\cdot f) = f$ for all $f \in \mathcal{D}(\Omega)$, where now U is
the integral operator defined by the kernel $\frac{1}{4\pi} G(x,y)$. In
both cases, $y \mapsto \frac{1}{|x-y|}$ and $y \mapsto G(x,y)$ are solutions of the
Laplace equation $\Delta u = 0$, but with a <u>singular point</u> for y=x.

In 1860 Riemann proposed to solve Cauchy's problem for linear hyperbolic equations of second order in 2 variables
$P\cdot u = 0$ by using a particular solution $y \mapsto R(x,y)$ of the
adjoint equation ${}^t P\cdot v = 0$ depending on the point x, in
such a way that the integral operator U defined by the kernel
R satisfies again the relation $P\cdot(U\cdot f) = f$ for $f \in \mathcal{D}(\mathbb{R}^2)$,
the function R playing for P a role similar to the Green
function for the laplacian. In that case R is continuous;
but when Volterra and Hadamard undertook to extend the method
to second order hyperbolic equations in $n \geq 3$ variables,
they were beset by difficulties stemming from the fact that
the function corresponding to R would now have, not only
(as the Green function) a singularity at one point x, but
singularities along lines or surfaces.

We cannot here describe the details of these researches,
for which we refer to [93]. What we want to emphasize is
that, by the end of the XIXth century, one had the (rather
vague) idea that, for a second order linear differential ope-

rator P ,   one should look for a solution $y \longmapsto R(x,y)$ of
the adjoint equation ${}^t F \cdot v = 0$ having a suitable singularity
at the point x ,   and that the integral operator U defined
by the kernel R would be such that $P \cdot (U \cdot f) = f$ for
$f \in \mathcal{D}(\Omega)$;   such a function R was called an elementary (or
fundamental) solution for ${}^t P$.   This worked, not only for Δ ,
but also for instance for the heat operator $\dfrac{\partial}{\partial t} - \dfrac{\partial^2}{\partial x^2}$,   one
has $R(t,x) = \dfrac{1}{2\sqrt{\pi\, t}}\, e^{-x^2/4t}$ for $t > 0$, $R(t,x) = 0$ for
$t < 0$ ($R(0,x)$ is undefined); for hyperbolic equations,
things were not so simple, for one had to apply the integral
operator defined by the kernel R ,   not to f but to some
derivatives of f [122].

   After 1900, mathematicians began to investigate the possibility of extending the notion of elementary solution and its
applications to equations of higher order. To an operator
(25), one associates the polynomial in $\xi = (\xi_1, \xi_2, \ldots, \xi_n)$

$$(28) \qquad\qquad \sigma_P(x,\xi) = \sum_{|\alpha| \le m} a_\alpha(x)(2\pi i \xi)^\alpha$$

(later called the "symbol" of P) and the homogeneous polynomial in ξ consisting in the terms of highest degree m
(the "principal symbol" of P)

$$(29) \qquad\qquad \sigma_P^0(x,\xi) = \sum_{|\alpha|=m} a_\alpha(x)(2\pi i \xi)^\alpha.$$

The operator P is called elliptic if $\sigma_P^0(x,\xi) \ne 0$ for all
$x \in \Omega$ and all $\xi \ne 0$. In his thesis, Fredholm considered,
for $n = 3$,   elliptic operators with constant coefficients,
and proved the existence of an elementary solution by writing
it explicitly as an abelian integral ([74], p.17-57); this was

later generalized to elliptic operators with constant coefficients in any number of variables (Holmgren, Herglotz [109]).
In 1907, E.E. Levi considered elliptic operators with variable coefficients, and either $n = 2$ variables, or operators
or order 2 in any number of variables; in both cases, using
the fact that for constant coefficients elementary solutions
were explicitly known, he showed how one could prove the existence of elementary solutions by showing that their determination could be reduced to the solution of a Fredholm integral
equation ([145], vol.II, p.28-84).

For operators with constant coefficients, the theory of distributions completely clarified the concept of "elementary
solution" [194]; such an operator may be written $u \mapsto A*u$,
where $A = \sum\limits_{\alpha} a_\alpha D^\alpha \varepsilon_o$, linear combination of derivatives of the
Dirac measure ε_o at the origin of $\mathbb{R}^n$. An elementary solution is then, by definition, a <u>distribution</u> E on $\mathbb{R}^n$ such
that

(30) $A*E = \varepsilon_o$;

it follows at once from that definition that, for any <u>distri-
bution</u> T with compact support (and not only for a function),
one has

(31) $A*(E*T) = E*(A*T) = T$.

In 1954, Ehrenpreis [63] and Malgrange [156] independently
proved that <u>any</u> operator P of form (25) with constant coefficients has an elementary solution E. Of course, such a
solution is only determined up to addition of any distribution S solution of the homogeneous equation $P \cdot S = 0$. It

is not obvious that among these distributions there would

exist <u>tempered</u> ones; as $\mathfrak{F}A$ is the polynomial $\sigma_p(\xi)$, such

an elementary solution should be such that

$$(32) \qquad\qquad\qquad \sigma_p(\xi) \cdot \mathfrak{F}E = 1;$$

it is only in 1958 that independently Hörmander [118] and

Łojasiewicz [153] showed that it is always possible, for any

polynomial Q, to find a tempered distribution T such that

$Q \cdot T = 1$.

Elementary solutions proved to be useful to show that an

operator P is hypoelliptic [117, p.100], or to prove uni-

queness of the Cauchy problem for some operators [117,p.141].

But gradually it was realized that instead of looking for a

"right inverse" to a differential operator, it could be much

simpler to obtain an "approximate right inverse" which would

be put to the same uses.

Such an idea was first introduced by Hilbert in 1907 under

the name of <u>parametrix</u>, in a particularly simple context, the

study of an elliptic operator P of order 2 on the sphere $\mathbb{S}_2$;

he proves that there is an integral operator Q, defined by

a kernel having a singularity similar to the logarithmic sin-

gularity of the Green function, and such that $QP = I + R$,

where R is an integral operator; a solution of $P \cdot u = f$

is therefore a solution of the <u>integral equation</u> $u + R \cdot u =$

$= Q \cdot f$ [112, p.233-242]; Q could therefore be considered as

an "approximate left inverse" of P.

Two years later, E.E. Levi independently introduced a simi-

lar method in a much more general and difficult question, the

generalization of the Dirichlet problem for elliptic operators

of arbitrary even order, completely unexplored until then
except for a few special operators such as the iterated laplacian. On these special examples it transpired that what
should correspond to the Dirichlet problem for an operator of
even order 2m was the boundary condition consisting in fix-
ing the values of the solution and its first m-1 normal de-
rivatives on the boundary Γ of a bounded open set $\Omega \subset \mathbb{R}^n$.
Levi is only concerned with the case of n = 2 variables and
first shows that the problem (for smooth boundary Γ) may be
reduced to the case in which one has to find a solution of
$P \cdot u = f$ such that u and its m-1 first normal derivatives
take the value 0 on Γ. His idea is then to determine two
functions, $\varphi(x,y)$ and $K(x,y,x_1,y_1)$ defined respectively
in $\bar{\Omega}$ and $\bar{\Omega} \times \bar{\Omega}$, and such that: 1º for each point $(x_1,y_1) \in$
$\in \Omega$, the function $(x,y) \mapsto K(x,y,x_1,y_1)$ and its m-1 first
normal derivatives vanish on Γ; 2º the function

$$(33) \qquad u(x,y) = \iint_\Omega K(x,y,x_1,y_1)\varphi(x_1,y_1)dx_1 dy_1$$

satisfies the equation $P \cdot u = f$. In order to obtain that
result, he chooses K in such a way that

$$(34) \qquad (P \cdot u)(x,y) = \varphi(x,y) + \iint_\Omega K_1(x,y,x_1,y_1)\varphi(x_1,y_1)dx_1 dy_1$$

where K_1 is a kernel to which Fredholm's theory is applicable, and he has thus reduced the problem to a Fredholm integral equation. If one writes $Q \cdot \varphi$ the right hand side of
(33), one may say that the operator Q is such that $PQ = I + R$,
where R is an integral operator; this time Q is an "approximate right inverse" of P. The determination of the func-

tion K satisfying these conditions is a difficult problem,
and it is not surprising that after Levi not much work was
done in that direction until around 1960 ([145], vol.II,
p. 207-343).

At that time progress came from a completely different direction. In his work on integrals of complex functions along
paths in $\mathbb{C}$, Cauchy, in 1814, had observed that if f is a
c^1 function in an interval [-a,a] of $\mathbb{R}$, the function
$f(x)/x$ is not integrable if $f(0) \neq 0$, but the sum
$\int_{-a}^{-\varepsilon} \frac{f(x)dx}{x} + \int_{\varepsilon}^{a} \frac{f(x)dx}{x}$, when ε tends to 0, has a limit
which he called the "principal value" of the integral. Simi-
larly, if L is a c^1 curve in C, the limit

$$(35) \qquad (H \cdot f)(x) = \lim_{\varepsilon \to 0} \int_{L_\varepsilon} \frac{f(z)dz}{z-x}$$

where $x \in L$ and L_ε is the part of L for which the arc
of L joining x and z has length $> \varepsilon$, exists for each
c^1 function f defined on L, and is written
$vp \int_L \frac{f(z)dz}{z-x}$; if L is a simple closed curve, the boundary
of a bounded open set Ω, the usual line integral
$\frac{1}{2\pi i} \int_L \frac{f(z)dz}{z-x}$ is defined for $x \notin L$, and is in Ω a holomorphic function $F^+(x)$, and in the exterior $\mathbb{C} - \bar{\Omega}$ another
holomorphic function $F^-(x)$; when x tends to a point $t \in L$,
these functions have limits respectively equal to

$$F^+(t) = \frac{1}{2} f(t) + \frac{1}{2\pi i} vp \int_L \frac{f(z)dz}{z-t} ,$$

$$F^-(t) = -\frac{1}{2} f(t) + \frac{1}{2\pi i} vp \int_L \frac{f(z)dz}{z-t} .$$

In his third paper on integral equations, Hilbert, using these formulas, showed that one could find two holomorphic functions $F^+(x)$ in Ω, $F^-(x)$ in $\mathbb{C} - \bar{\Omega}$, such that for $t \in L$, the limits $F^+(t)$ and $F^-(t)$ of these functions exist when x tends to t, and satisfy a relation $F^+(t) = $ $= g(t)F^-(t)$, where g is a C^1 function on L [112, p. 81-108]; this led to calling the function $H \cdot f$ defined by (35) the <u>Hilbert transform</u> of f.

Between 1910 and 1955, many mathematicians studied various generalizations of this operator to functions of any number of variables, and applied them to various problems of Analysis; we cannot describe this evolution in any detail, and refer the reader to [197]. The most general of these "singular integral operators" (or "Calderon-Zygmund operators" as they were also called) are defined in the following way: Ω is an open subset of $\mathbb{R}^n$, $(x,\xi) \mapsto K(x,\xi)$ a locally integrable mapping of $\Omega \times (\mathbb{R}^n - \{0\})$ into $\mathbb{C}$, which is <u>positively homogeneous of degree</u> $-n$ in ξ for every $x \in \Omega$; in addition it is assumed that for any $x \in \Omega$, $\int_{S_{n-1}} K(x,\xi)d\sigma(\xi) = 0$ (σ being the invariant measure on S_{n-1}). Then, for any function $u \in \mathcal{D}(\Omega)$, the limit

$$(36) \qquad (P \cdot u)(x) = \lim_{\varepsilon \to 0} \int_{|y-x| \geq \varepsilon} K(x,y-x)u(y)dy$$

exists for every x and defines a <u>singular integral operator</u> P.

Around 1960, it was realized that the use of Fourier transform (generalized to distributions) enabled one to define a class of linear operators which contained at the same time

differential operators of type (25), singular integral operators and some ordinary integral operators (with locally integrable kernels); several mathematicians independently contributed to this new theory, but again we cannot go into any
historical detail, and we shall merely give a short description of its present status (for more references, see [D],
[119] and [68]). The inversion formula for Fourier transforms shows that the operator (25) can be written

$$(37) \quad (P \cdot u)(x) = \int_{\mathbb{R}^n} \exp(2\pi i(x|\xi))a(x,\xi)\mathcal{F}u(\xi)d\xi \quad \text{for} \quad u \in \mathcal{D}(\Omega)$$

where $a(x,\xi) = \sigma_P(x,\xi)$ (formula (28)). The generalization
consists in replacing in (37) the polynomial (in ξ) $\sigma_P(x,\xi)$
by a more general C^∞ function defined in $\Omega \times \mathbb{R}^n$ and which
is only submitted to conditions concerning its <u>growth</u> as $|\xi|$
tends to $+\infty$: for a polynomial σ_P, $D_x^\alpha \sigma_P$ has the same be-
havior for $|\xi| \to \infty$ as σ_P itself, whereas $D_\xi^\beta \sigma_P$ is a poly-
nomial in ξ of degree $m-|\beta|$. One defines then a <u>symbol</u> as
a C^∞ mapping $(x,\xi) \mapsto a(x,\xi)$ of $\Omega \times \mathbb{R}^n$ into $\mathbb{C}$, such that
one has

$$(38) \qquad\qquad |D_x^\alpha D_\xi^\beta a(x,\xi)| \leq C_{\alpha\beta L} (1+|\xi|)^{m-|\beta|}$$

for all multiindices α, β and all compact subsets $L \subset \Omega$,
where $x \in L$ and $\xi \in \mathbb{R}^n$ are arbitrary, and $C_{\alpha\beta L}$ is independent of $x \in L$ and $\xi \in \mathbb{R}^n$; the main difference is that
here m (the <u>order</u> of a, or of P) is an <u>arbitrary real</u>
number. The corresponding operator P defined by (37) for
$u \in \mathcal{D}(\Omega)$ is called a <u>pseudo-differential operator</u> defined
by the symbol a; differential operators therefore correspond
to symbols of order m equal to an integer $m \geq 1$, the new-

tonian potential to $m = -2$,   and the singular integral ope-

rators (36) (for   K   of class C^∞)   to $m = 0$.   The most in-

teresting case is the one in which $a = a_o + a_1$,   where a_o is po-

sitively homogeneous of degree   m   in ξ ,   and a_1 is a

symbol of order $< m$;   one then says that a_o is the princi-

pal symbol of the pseudo-differential operator (37), and one

writes $a_o = \sigma_P^o$.

The main properties of pseudo-differential operators are

the following ones:

I) P maps the space $\mathcal{D}(\Omega)$ into the space $\mathcal{E}(\Omega)$ of all C^∞

complex functions in Ω;   it has an adjoint P^*,   satisfying

$$(39) \qquad\qquad (P \cdot u \,|\, v) = (u \,|\, P^* \cdot v)$$

for all   u, v   in $\mathcal{D}(\Omega)$ (scalar product of $L^2(\Omega)$),   which

is a pseudo-differential operator of same order   m;   if P

has a principal symbol, P^* has a principal symbol such that

$$(40) \qquad\qquad \sigma_{P*}^o = \overline{\sigma_P^o}.$$

II) One says P is of proper type if both P and P^* apply

$\mathcal{D}(\Omega)$ into itself;   for any pseudo-differential operator Q

of order   r,   the compositions QP and PQ are then defined

and are pseudo-differential operators of order   m+r;   if P

and Q have principal symbols, so have PQ and QP and

$$(41) \qquad\qquad \sigma_{PQ}^o = \sigma_{QP}^o = \sigma_P^o \sigma_Q^o \ .$$

III) If $m < -1$, P is an integral operator, having a

kernel which is locally integrable in $\Omega \times \Omega$,   but has singular-

ities for $x = y$;   if $m < -n-k$,   the kernel is of class C^k

in the whole of $\Omega \times \Omega$.   One says that the symbol a (and the

corresponding pseudo-differential operator P) are of order
-∞ if a satisfies inequalities (38) for every real number m;
P is then an integral operator with a kernel which is of
class C^∞, and conversely any such operator is a pseudo-differential operator of order -∞, and its principal symbol
is 0.

IV) Any pseudo-differential operator P may be extended by
continuity (for the weak topology) from $\mathcal{D}(\Omega)$ to the space
$\mathcal{E}'(\Omega)$ of all distributions on Ω with compact support. The
operators P of order -∞ are characterized by the property
that for any distribution $T \in \mathcal{E}'(\Omega)$, $P \cdot T$ is a C^∞ function
on Ω; one says that these operators are smoothing operators.
Any pseudo-differential operator is the sum of a pseudo-differential operator of proper type and of a smoothing operator.
When K is a smoothing operator, so are the products KP and
PK for any pseudo-differential operator P, if one of the
two operators K , P is of proper type.

One writes $P \sim Q$ if $P - Q$ is a smoothing operator.

V) The most remarkable feature of pseudo-differential operators is the possibility of defining a symbol by an asymptotic
expansion. Suppose given an infinite sequence $a_0, a_1, \ldots, a_k$,..
.., of symbols, having orders $m_0 > m_1 > \ldots > m_k > \ldots$ with
$\lim_{k \to \infty} m_k = -\infty$; then there exists a symbol a of order m_0
such that, for any k, $a - (a_0 + a_1 + \ldots + a_k)$ has order $< m_k$;
this is expressed by writing

$$a \sim a_0 + a_1 + \ldots + a_k + \ldots$$

and saying that the right hand side is an asymptotic expansion
of a. If $P_0, \ldots, P_k, \ldots$ are the pseudo-differential opera-

tors defined by the symbols $a_o, \ldots, a_k, \ldots,$ and P the
pseudo-differential operator defined by a, one also writes

$$P \sim P_o + P_1 + \ldots + P_k + \ldots$$

VI) One says a pseudo-differential operator P having a
principal symbol of order m is <u>elliptic</u> if $\sigma_P^o(x,\xi) \neq 0$
for $x \in \Omega$ and $\xi \neq 0$; for differential operators, this
coincides with the previous definition. It is <u>equivalent</u> to
say that there exists a pseudo-differential operator Q of
order -m and of proper type such that $QP = I+R$ and $PQ = I+R'$ where
R and R' are smoothing operators; in other words, P has a
(left and right) <u>parametrix</u> in a very strong sense. The proof
is very simple; the necessity follows from the fact that if
$QP \sim I$, one must have $\sigma_P^o = (\sigma_Q^o)^{-1}$ by (41). Conversely,
if P is elliptic, there is a pseudo-differential operator Q_1
of proper type defined by the symbol $(\sigma_P^o)^{-1}$; one has then
$Q_1 P = I - P_1$, where P_1 has <u>order</u> $\leq -1,$ and one is reduced
to finding a pseudo-differential operator Q_2 such that
$Q_2(I-P_1) = I+R$, where R is a smoothing operator; but it
is enough to take $Q_2 \sim I + P_1 + P_1^2 + \ldots + P_1^k + \ldots$ to obtain
that result!

VII) An immediate consequence of the existence of a parametrix Q for an elliptic differential operator P is that
P is <u>hypoelliptic</u>, for if T is a distribution such that
$P \cdot T = f \in \mathcal{E}(\Omega)$, one has $Q \cdot f = T + R \cdot T$, and as R is a
smoothing operator, $R \cdot T$ and $Q \cdot f$ are both C^∞ functions,
hence also T. Another easy consequence of the use of pseudo-
differential operators is that for each point $x_o \in \Omega$, there
is a small neighborhood $U \subset \Omega$ of x_o such that the equation

$P \cdot u = f$ has solutions in U (in other words, the H. Lewy phenomenon (chap.II, §2) cannot occur for an elliptic differential operator P); one should note, however, that there are examples of elliptic operators P defined in $\mathbb{R}^n$, and such that there are equations $P \cdot u = f$ which have <u>no solution</u> in a large ball containing the support of $f \in \mathcal{O}(\mathbb{R}^n)$ [176].

VIII) When $\Omega \subset \mathbb{R}^n$ is bounded, pseudo-differential operators of proper type in Ω have simple continuity properties with respect to the Sobolev spaces: if P is such an operator of order r , defined in a neighborhood of $\bar{\Omega}$, and s is any real number, there is a constant C depending only on P and s, such that

(42) $\| P \cdot u \|_s \leq C \| u \|_{r+s}$

for any $u \in \mathcal{O}(\Omega)$, the norms being those of $H^s(\mathbb{R}^n)$ and $H^{r+s}(\mathbb{R}^n)$. If $r > 0$ and P is <u>elliptic</u>, applying this result to a parametrix of P immediately yields an <u>a priori</u> inequality of Friedrichs type

(43) $\| u \|_r \leq C(\| P \cdot u \|_o + \| u \|_o)$.

Suppose in addition that $P = P^*$ is a differential operator such that $\sigma_P(x, \xi) > 0$ for large $|\xi|$; then an easy inductive argument determines (by an asymptotic expansion of its symbol) a pseudo-differential operator S of proper type and of order $r/2$ such that $P = S^* S + R$, where R is a smoothing operator. Applying (42) and (43) to S , one obtains the existence of constants $a > 0$, $b > 0$, $c > 0$ such that, for u and v in $\mathcal{O}(\Omega)$, one has

(44) $| (P \cdot u | v)_o | \leq c \| u \|_{r/2} \| v \|_{r/2}$

(45) $(P \cdot u | u)_0 \geq a \|u\|^2_{r/2} - b\|u\|^2_0$.

The second one (for an even integer r) was first proved by
Gårding in 1953 [80]. It enabled him to apply the von Neumann
spectral theory to the hermitian operator T_P in the Hilbert
space $L^2(\Omega)$, with $\text{dom}(T_P) = \mathcal{D}(\Omega)$. In general, the defects
of T_P^{**} are both infinite, and one can define a particular
<u>self-adjoint</u> extension A_P of T_P by the following process:
$\text{dom}(A_P)$ is the dense subspace of $L^2(\Omega)$ consisting of functions u such that the <u>distribution</u> $P \cdot u$ is again in $L^2(\Omega)$
and then $A_P \cdot u = P \cdot u$; furthermore, $\text{dom}(A_P)$ is contained in
the space $H_0^{r/2}(\Omega)$, the closure in $H^{r/2}(\mathbb{R}^n)$ of $\mathcal{D}(\Omega)$. The
spectrum of A_P is reduced to the point spectrum, consisting
of an increasing sequence (λ_n) of real eigenvalues of finite multiplicity, tending to $+\infty$; the corresponding eigenfunctions (suitably normalized) form a <u>Hilbert basis</u> of $L^2(\Omega)$
and are of class C^∞; for any $\zeta \in \mathbb{C}$ distinct from the λ_n,
$G_\zeta = (A_P - \zeta I)^{-1}$ is a <u>compact</u> operator, which one may call the
<u>Green operator</u> of $P - \zeta I$. It is easy to see that the res-
triction of G_ζ to $\mathcal{D}(\Omega)$ is a pseudo-differential operator
of order $-r$, which in general is <u>not</u> of proper type; however, for every distribution T with compact support in Ω,
one has $(P-\zeta I) \cdot (G_\zeta \cdot T) = G_\zeta \cdot ((P-\zeta I) \cdot T) = T$ and in particular,
for any point $x \in \Omega$,

(46) $(P-\zeta I) \cdot (G_\zeta \cdot \varepsilon_x) = G_\zeta \cdot ((P-\zeta I) \cdot \varepsilon_x) = \varepsilon_x$

so that one may say that the distribution $G_\zeta \cdot \varepsilon_x$ is an <u>ele-
mentary solution</u> of $P - \zeta I$ at the point x.

IX) The results of VIII) apply in particular to a differential operator of even order $2p \geq 2$

$$(47) \qquad (P \cdot u)(x) = \sum_{|\alpha| \leq p, |\beta| \leq p} D^{\alpha}(a_{\alpha\beta}(x)D^{\beta}u(x))$$

where the $a_{\alpha\beta}$ are <u>bounded</u> C^{∞} functions in a neighborhood
of the <u>bounded</u> set $\bar{\Omega}$,   such that:

1° $(-1)^{|\beta|} a_{\beta\alpha} = (-1)^{|\alpha|} \overline{a_{\alpha\beta}}$,   which guarantee that $P^* = P$;

2°   there is a constant $C > 0$ such that, for every $x \in \Omega$
and every family $(z_{\alpha})_{|\alpha|=p}$ of complex numbers, one has

$$(48) \qquad \sum_{|\alpha|=|\beta|=p} (-1)^p a_{\alpha\beta}(x) z_{\alpha} \overline{z_{\beta}} \geq C(\sum_{|\alpha|=p} |z_{\alpha}|^2).$$

The Green operator G_{ζ} is then (for $\zeta \in Sp(A_p)$)   an <u>inte-</u>
<u>gral operator</u>,   with kernel $(x,y) \mapsto G(\zeta,x,y)$ which is local-
ly integrable in $\Omega \times \Omega$, C^{∞} outside of the diagonal and such
that $G(\bar{\zeta},x,y) = \overline{G(\zeta,y,x)}$ (the <u>Green function</u> of P); from
(46) it follows that $y \mapsto G(\zeta,x,y)$ is a <u>solution</u> of $P \cdot u =$
$= \zeta u$ in the complement $\Omega - \{x\}$ of the point   x.

One may always take $\zeta = -b$ for a sufficiently large num-
ber $b > 0$,   and for every function $f \in \mathcal{E}(\Omega) \cap L^2(\Omega)$,   there
is therefore <u>a unique solution of the equation</u> $P \cdot u + bu = f$
<u>belonging to the space</u> $H_o^p(\Omega)$ <u>and of class</u> C^{∞} <u>in</u> Ω.

These results were obtained (of course without the theory
of pseudo-differential operators) by Gårding and Višik (independently) in 1953 ([80], [218]);   they may be considered as
a "weak" solution of the generalization of Dirichlet's problem
considered by E.E. Levi: <u>no</u> assumption is made on the
boundary Γ of Ω,   but all which is required of the solution
is that it should be arbitrarily close, <u>for the topology of</u>

$H^p(\mathbb{R}^n)$, of C^∞ functions vanishing in a neighborhood to Γ; but it (or its derivatives) may have a very pathological behaviour at points of Γ if Γ is not smooth.

If one makes the additional assumption that

$$(P \cdot u)(x) = \sum_{|\alpha|=p, |\beta|=p} D^\alpha (a_{\alpha\beta}(x) D^\beta u(x)) + \sum_{|\nu|<p} D^\nu (a_\nu(x) D^\nu u(x))$$

where $(-1)^{|\nu|} a_\nu(x) \geq 0$ for all $|\nu| < p$, then one may even take $b = 0$ in the Gårding-Višik theorem (this is the case in particular for $(-\Delta)^p$).

X) If E and F are two complex vector bundles over a compact differentiable manifold X, and $\Gamma(E)$, $\Gamma(F)$ are the vector spaces of C^∞ sections of these bundles over X, one can define pseudo-differential operators $P: \Gamma(E) \to \Gamma(F)$, which become matrices of ordinary pseudo-differential operators when expressed in local coordinates. It is then possible to define intrinsically a <u>principal symbol</u> σ_P^0: for each $x \in X$ and every tangent covector ξ to X at the point x, $\sigma_P^0(x,\xi)$ is a homomorphism $E_x \to F_x$ of the vector spaces, fibres of E and F at x; in local coordinates, $\sigma_P^0(x,\xi)$ is the matrix of the principal symbols of the elements of the matrix equal to P. It is possible to define on $\Gamma(E)$ and $\Gamma(F)$ structures of prehilbert spaces, and to attach to any pseudo-differential operator P its adjoint $P^*: \Gamma(F) \to \Gamma(E)$ such that $(P \cdot u | v) = (u | P^* \cdot v)$; properties (40) and (41) still hold.

A pseudo-differential operator $P: \Gamma(E) \to \Gamma(E)$ is then called <u>elliptic</u> if for every $x \in X$ and every $\xi \neq 0$, $\sigma_P^0(x,\xi)$ is a <u>bijection</u> of E_x onto itself, and the existence

of a <u>parametrix</u> for such an operator can then be proved as in
VI). For elliptic operators such that $P^* = P$, the application of spectral theory to the hermitian operator T_P (in
the Hilbert space, completion of $\Gamma(E)$) is here much simpler
than in VIII) due to the absence of "boundary conditions":
there is a Hilbert basis (u_k) of $\Gamma(E)$ such that $P \cdot u_k =$
$= \mu_k u_k$, where μ_k is real and $|\mu_k|$ tends to $+\infty$ with k;
for every $f \in \Gamma(E)$, one has $f = \sum_k (f|u_k)u_k$, the series
being convergent for the topology of the Fréchet space $\Gamma(E)$,
and $P \cdot f = \sum_k \mu_k(f|u_k)u_k$ with the same convergence. In par-
ticular, $\mathrm{Ker}(P)$ is the finite dimensional subspace having
as a basis the u_k for which $\mu_k = 0$, and $\mathrm{Im}(P)$ is closed
and is a topological supplement of $\mathrm{Ker}(P)$.

If now P is <u>any</u> elliptic operator $\Gamma(E) \to \Gamma(E)$, P^*P and
PP^* are both elliptic and equal to their adjoints; the study
of these operators enables one to evaluate the difference
$\dim(\mathrm{Ker}(P)) - \dim(\mathrm{Ker}(P^*))$, the <u>index</u> of P, and to express
it by a formula in terms of the principal symbol of P and of
the cohomology of X; this is the famous <u>Atiyah-Singer for-
mula</u>, a fundamental result which has many applications and
has spurred research in many directions ([10], [31], [198],
[199]). It was in fact due to the development of the necessary tools for the proof of that formula that the theory of
pseudo-differential operators got started.

XI) The Gårding-Višik theorem of IX) leaves unanswered two
questions: 1º Why is it that, in the Dirichlet problem and
its generalizations, <u>half</u> of the Cauchy data on the boundary
are enough to determine the solution? 2º What can be said of

the behavior of the unique solution belonging to $H_o^p(\Omega)$ in the vicinity of a point of the boundary Γ where Γ is smooth?

To answer these questions, one starts by investigating the Cauchy problem for an elliptic operator P and seeing why in general it has no solution. Suppose that the bounded open set $\Omega \subset \mathbb{R}^n$ has a smooth boundary Γ, and (for simplicity's sake) that P is a differential operator defined in a neighborhood Ω_o of $\bar{\Omega}$, has even order $2p \geq 2$, and possesses in Ω_o a parametrix Q of order $-2p$ such that $Q \cdot (P \cdot T) = P \cdot (Q \cdot T)$ for any distribution $T \in \mathcal{E}'(\Omega_o)$ (this is the case for the operator $P - \zeta I$ in VIII), but here we do not suppose that $P^* = P$). For any function $u \in \mathcal{E}(\Omega_o)$, we note $\mathrm{Dch}(u)$ the function $(g_o, g_1, \ldots, g_{2p-1})$ defined on the boundary Γ and with values in $\mathbb{C}^{2p}$, where g_j is the normal derivative $\dfrac{\partial^j u}{\partial n^j}$ at a point of Γ. The starting point is an idea due to Sobolev ([200], p.63): let u^o be the discontinuous function equal to u in $\bar{\Omega}$ but to 0 in the complement $\Omega_o - \bar{\Omega}$. Then $P \cdot u^o$ is well defined as a distribution on Ω_o, and it is easy to check that one can write

$$(49) \qquad P \cdot u^o = (P \cdot u)^o + N \cdot \mathrm{Dch}(u)$$

where N is a linear operator (independent of the function u) which to every C^∞ function in $(\mathcal{E}(\Gamma))^{2p}$ associates a distribution with support in Γ (what one now calls a multilayer on Γ). As both sides of (49) are distributions on Ω_o with compact support, the operator Q may be applied to them, and yields the relation

(50) $u^o = Q \cdot ((P \cdot u)^o) + Q \cdot (N \cdot \text{Dch}(u)).$

This is the general form of Green's formula (17) of chap. II,
§3; for any function $f \in \mathcal{e}(\Omega_o)$, the distribution $Q \cdot f^o$ has
a restriction to Ω which is a C^∞ <u>function</u> such that all
its derivatives have <u>limits</u> at every point of Γ; one says
that it is the Q-<u>potential of the mass distribution</u> of den-
sity f on Ω. Similarly, for any vector function $g \in (\mathcal{e}(\Gamma))^{2p}$,
the distribution $Q \cdot (N \cdot g)$ has the same properties, and one
says that its restriction to Ω is the Q-<u>potential of the</u>
<u>multilayer</u> $N \cdot g$; these properties obviously generalize the
classical properties of the newtonian potentials of a mass
distribution, of a single layer and of a double layer (chap.
II, §3) (of course the restriction of $Q \cdot (N \cdot g)$ to $\Omega_o - \bar{\Omega}$
also has similar properties, but the limits at a point of Γ
differ in general from the limits at the same point of the
restriction of $Q \cdot (N \cdot g)$ to Ω). Therefore, equation (50)
shows that if there is an $u \in \mathcal{e}(\Omega)$ such that $P \cdot u = f$ and
$\text{Dch}(u) = g$ are given functions, this function u (restriction of u^o) is <u>unique</u>, which corresponds to what one may
expect of the Cauchy problem. But in addition one must have
$\text{Dch}(u^o) = g$, which gives the <u>necessary condition</u>

(51) $g = \text{Dch}(Q \cdot f^o) + \text{Dch}(Q \cdot (N \cdot g))$

between f and g. One proves that $C : g \mapsto \text{Dch}(Q \cdot (N \cdot g))$
is a pseudo-differential operator of $(\mathcal{e}(\Gamma))^{2p}$ into itself,
which is called the <u>Calderon operator</u> corresponding to the
parametrix Q.

A more detailed study shows that (51) is equivalent to p

linear relations between the 2p functions $g_o, \ldots, g_{2p-1}$ and their derivatives; this explains why one cannot prescribe the 2p functions u, $\frac{\partial u}{\partial n}, \ldots, \frac{\partial^{2p-1}u}{\partial n^{2p-1}}$ on Γ, but only p of them. More generally, one may consider a differential operator B of $(\mathcal{E}(\Gamma))^{2p}$ into $(\mathcal{E}(\Gamma))^p$, and instead of the Cauchy problem, consider the boundary conditions $B \cdot (\mathrm{Dch}(u)) = g$ for a given vector function $g \in (\mathcal{E}(\Gamma))^p$. It is then possible to describe explicitly a set of sufficient conditions (called the <u>Lopatinski conditions</u>) linking B and C and implying that the problem can be reduced to Fredholm integral equations on Γ; more precisely, these conditions imply that the mapping

$$(52) \qquad\qquad u \longmapsto (P \cdot u, B \cdot (\mathrm{Dch}(u)))$$

of $\mathcal{E}(\bar{\Omega})$ (space of the restrictions to $\bar{\Omega}$ of functions of $\mathcal{E}(\Omega_o)$) into $\mathcal{E}(\bar{\Omega}) \times (\mathcal{E}(\Gamma))^p$ has <u>a finite dimensional kernel and a closed image of finite codimension.</u>

In particular, one checks that the Lopatinski conditions are always satisfied if one takes for $B \cdot g$ a consecutive sequence $(g_q, g_{q+1}, \ldots, g_{q+p-1})$ of p of the functions g_j, and the corresponding problem for q = 0 is just the Dirichlet problem as posed by E.E. Levi.

At this point, one might think, from the example (47), that except for a <u>denumerable</u> set of values of $\zeta \in \mathbb{C}$, the mapping $u \longmapsto ((P - \zeta I) \cdot u, B \cdot (\mathrm{Dch}(u)))$ would in fact be <u>bijective</u>. However, this is <u>not</u> always the case, and there are examples for which that mapping is injective for <u>no</u> $\zeta \in \mathbb{C}$.

For operators (47) to which the Gårding-Višik theorem applies, to prove that the preceding mapping is bijective for $\zeta \in \mathrm{Sp}(A_p)$

(with $B \cdot g = (g_o, \ldots, g_{p-1})$), it is enough to show that, when Γ is smooth, the unique solution $u \in H^p_o(\Omega)$, and all its derivatives, can be extended by continuity to $\bar{\Omega} = \Omega \cup \Gamma$ (the second problem mentioned above). Actually, even if Γ is not smooth everywhere, the existence of limits for these functions is guaranteed at <u>each point</u> where Γ is smooth; this was first proved by L. Nirenberg in 1955 [167], and has been proved by Peetre in 1961 using a different method, which however still relies on <u>a priori</u> inequalities [171]. These results may be extended to other types of boundary conditions $B \cdot (Dch(u)) = g$ satisfying the Lopatinski conditions, the so-called <u>coercitive</u> problems for elliptic operators P for which $P^* = P$.

<u>Further generalizations</u>. Formula (37) defining a pseudo-differential operator can also be written, replacing $\mathfrak{F}u$ by its definition

$$(P \cdot u)(x) = \widetilde{\iint_{\Omega \times \mathbb{R}^n}} \exp(2\pi i(x-y|\xi))a(x,\xi)u(y)dyd\xi$$

where the integral is not any more a Lebesgue integral, but an "improper" (or "oscillating") one, obtained by passage to the limit from the integral of the same function multiplied by a function $h(\xi/q)$, where $h \in \mathcal{D}(\mathbb{R}^n)$ is equal to 1 in a neighborhood of 0, and q tends to $+\infty$. It turns out that one can define similar integrals when one replaces $\exp(2\pi i(x-y|\xi))$ by "phase functions" $\varphi(x,y,\xi)$ positively homogeneous in ξ, and $a(x,\xi)$ by more general "symbols" $a(x,y,\xi)$.

I) Such operators naturally occur in the theory of <u>strictly</u>

hyperbolic operators, of which the simplest is the wave operator (or dalembertian)

$$(53) \qquad \Box u = \frac{\partial^2 u}{\partial t^2} - (\frac{\partial^2 u}{\partial x_1^2} + \ldots + \frac{\partial^2 u}{\partial x_n^2}) \; .$$

The Cauchy problem for that operator, consisting in finding a
solution of $\Box u = 0$ such that $u(0,x) = g_0(x)$ and
$\frac{\partial u}{\partial t}(0,x) = g_1(x)$ are given functions, had already been solved
by Cauchy; the explicit formula he gave for the solution can
be written $u(t,x) = u_+(t,x) + u_-(t,x)$, where

$$(54) \; u_\pm (t,x) = \frac{1}{2} \iint exp(2\pi i((x-y|\xi)\pm|\xi|t)))(g_0(y) \pm \frac{1}{2\pi i} \frac{g_1(y)}{|\xi|})dy d\xi$$

the integrals being "improper" in a sense easy to describe.
In general, one considers an operator of order m in $n+1$
variables $t, x_1, \ldots, x_n$

$$(55) \qquad P \cdot u = \frac{\partial^m u}{\partial t^m} + \sum_{j=1}^{m} \sum_{|\alpha| \le j} c_{j\alpha}(t,x) \; D_x^\alpha (\frac{\partial^{m-j} u}{\partial t^{m-j}})$$

and one assumes that its principal symbol $\sigma_P(\tau,\xi,t,x)$ can
be written

$$(56) \qquad \sigma_P^0(\tau,\xi,t,x) = \prod_{j=0}^{m-1} (\tau - q_j(t,x,\xi))$$

where the q_j are real functions of class C^∞ in $I \times \Omega \times (\mathbb{R}^n - \{0\})$
(I open subset of $\mathbb{R}$, Ω open subset of $\mathbb{R}^n$), positively
homogeneous or degree 1 in ξ, and such that for $j \ne k$,
$q_j(t,x,\xi) \ne q_k(t,x,\xi)$ everywhere. The Cauchy problem to be
solved is to find a function $v(t,x)$ such that $P \cdot v = f$ and
$\frac{\partial^j}{\partial t^j} v(t_0,x) = g_j(x)$ $(0 \le j \le m-1)$ for a $t_0 \in I$, in a convenient neighborhood of $(t_0,x_0) \in I \times \Omega$, f and g_j being C^∞
functions. Taking (54) as a model, one introduces m^2 opera-

tors $E_{jh}(s)$ $(0 \le j,h \le m-1)$ for s in a neighborhood of

t_o in I, such that, if $E_h(s) = \sum\limits_{j=0}^{m-1} E_{jh}(s)$ for $0 \le h \le m-1$,

one has (locally)

(57) $PE_h(s) = R_h(s)$, $(\frac{\partial}{\partial t})^k E_h(s) = \delta_{hk} I$ for t=s and $0 \le k \le m-1$,

where the $R_h(s)$ are <u>smoothing</u> operators. If one writes

$$(L \cdot u)(t,x) = \int_{t_o}^{t} (E_{m-1}(s) \cdot u(s,\cdot))(t,x)ds$$

one has $(\frac{\partial}{\partial t})^k (L \cdot u)(t_o,x) = 0$ for $0 \le k \le m-1$, and

$P \cdot (L \cdot u) = u - V \cdot u$, where V is a <u>Volterra integral operator</u>

(58) $$(V \cdot u)(t,x) = \int_{t_o}^{t} ds \int_{U} K(t,s,x,y)u(s,y)dy$$

U being a neighborhood of x_o in Ω and K a C^∞ function.

It is easy to see that $I+V$ is inverted by $I+W$, where W

is a similar Volterra operator. The Cauchy problem is then

solved by taking in a sufficiently small neighborhood of

(t_o,x_o)

(59) $v = \sum\limits_{j=0}^{m-1} E_j(t_o) \cdot g_j + L \cdot ((I+W) \cdot (f - \sum\limits_{j=0}^{m-1} R_j(t_o) \cdot g_j))$.

The construction of the E_h follows an idea introduced by

P. Lax in 1957, and patterned after the known behavior of the

solutions of the wave equation, which "propagate" along "rays".

For operators (25) with analytic coefficients, it follows from

the Cauchy-Kowalewska theorem that the Cauchy problem for data

given on a hypersurface will fail to have a unique solution if

the hypersurface is given locally by an equation $z(x_1,...,x_n) =$

= const., where z is a solution of the partial differential

equation of order 1

(60) $$\Phi(x_1,\ldots,x_n, \frac{\partial z}{\partial x_1},\ldots,\frac{\partial z}{\partial x_n}) = 0$$

in which Φ is obtained from the principal symbol $\sigma_P^o(x,\xi)$ by replacing the vector ξ by $(\frac{\partial z}{\partial x_1},\ldots,\frac{\partial z}{\partial x_n})$; such hyper-surfaces are called <u>characteristic</u> for the operator P.

For strictly hyperbolic operators (55), it follows from (56) that the equation of characteristic hypersurfaces (60) splits into m equations

(61) $$\frac{\partial z}{\partial t} - q_j(t,x,\mathrm{grad}_x z) = 0, \qquad 0 \le j \le m-1$$

(with $\mathrm{grad}_x z = (\frac{\partial z}{\partial x_1},\ldots,\frac{\partial z}{\partial x_n})$). For the wave operator (53) the equations (61) are

$$\frac{\partial z}{\partial t} = \pm |\mathrm{grad}_x z|$$

with solutions

$$z = 2\pi((x|\xi) \pm |\xi|t)$$

which reduce to $2\pi(x|\xi)$ for $t = 0$; they are precisely the "phase functions" which enter in Cauchy's formula (54). For general strictly hyperbolic operators (54), one therefore introduces the m^2 operators $F_{jh}(s)$ defined by

$$(F_{jh}(s)\cdot u)(t,x) =$$

(62)

$$= \iint_{U \times \mathbb{R}^n} \exp(i(\psi_j(t,s,x,\xi)-2\pi(y|\xi)))a_{jh}(t,s,x,y,\xi)u(y)\,dy\,d\xi$$

where $\psi_j(t,s,x,\xi)$ is the unique solution of (61) satisfying the initial condition

(63) $$\psi_j(s,s,x,\xi) = 2\pi(x|\xi)$$

and a_{jh} is a symbol of order $-h$ (in the sense defined for pseudo-differential operators). The goal is to determine the a_{jh} in such a way that, if one writes $F_h(s) = \sum\limits_{j=0}^{m-1} F_{jh}(s)$, the following conditions are satisfied:

1º $Q_h(s) = P \circ F_h(s)$ is a smoothing operator;

2º for each $g \in \mathcal{D}(U)$, the restriction to the hyperplane $t = s$ of the function $\dfrac{\partial^k}{\partial t^k} (F_h(s) \cdot g)$ has the form $(s,x) \mapsto \delta_{hk} g(x) + (Q_{hk}(s) \cdot g)(x)$ where $Q_{hk}(s)$ is a smoothing operator, for $0 \leq h, k \leq m-1$.

The conditions (57) are then met by taking

$$(R_h(s) \cdot g)(t,x) = \sum_{k=0}^{m-1} \frac{(t-s)^k}{k!} (Q_{hk}(s) \cdot g)(x)$$

and
$$E_h(s) = F_h(s) - R_h(s).$$

To achieve that goal, one defines the a_{jh} by __asymptotic expansions__

$$(64) \qquad\qquad a_{jh} \sim \sum_{\ell=0}^{\infty} a_{jh}^{(\ell)}$$

where $a_{jh}^{(\ell)}$ is a symbol or order $-h - \ell$; the $a_{jh}^{(\ell)}$ are determined by __induction on__ ℓ, in such a way that if one writes

$$(F_{jhN}(s) \cdot u)(t,x) =$$

$$= \sum_{\ell=0}^{N} \iint_{U \times \mathbb{R}^n} \exp(i(\psi_j(t,s,x,\xi) - 2\pi(y|\xi))) a_{jh}^{(\ell)}(t,s,x,y,\xi) u(y) dy d\xi$$

then:

1º Each operator $P \circ F_{jhN}(s)$ is defined by a symbol of order $m-h-N-2$.

2º If $F_{hN}(s) = \sum\limits_{j=0}^{m-1} F_{jhN}(s)$, the __restriction to__ $t = s$ of the function $\dfrac{\partial^k}{\partial t^k}(F_{hN}(s) \cdot g)$ has the form $(s,x) \mapsto \delta_{hk} g(x) +$

$+ (Q_{hkN}(s) \cdot g)(x)$,   where $Q_{hkN}(s)$ is a pseudo-differential operator of order   k-h-N-1,   for $0 \le h, k \le m-1$.

It is in this inductive process that the analogs of the "rays" enter. The classical Cauchy method for integration of partial differential equations of order   1   consists, for each equation   (61),   in considering, in the space $I \times \Omega \times \mathbb{R}^{n+1}$,   the "characteristic" curves

$$t \mapsto (t, x_1(t), \ldots, x_n(t), p_0(t), p_1(t), \ldots, p_n(t))$$

solutions of the system of ordinary differential equations

$$(65) \quad \begin{cases} \dfrac{dx_k}{dt} = - \dfrac{\partial q_j}{\partial \xi_k} (t, x_1, \ldots, x_n, p_1, \ldots, p_n) \\[3mm] \dfrac{dp_k}{dt} = \dfrac{\partial q_j}{\partial x_k} (t, x_1, \ldots, x_n, p_1, \ldots, p_n) \end{cases} \quad 1 \le k \le n$$

and verifying the condition $p_0 = q_j(t, x_1, \ldots, x_n, p_1, \ldots, p_n)$; one says that their projections $t \mapsto (t, x_1(t), \ldots, x_n(t))$ on $I \times \Omega$ constitute the j-th family of <u>bicharacteristic curves</u> for the operator P.   For the wave operator, the bicharacteristic curves which are such that $x_k(s) = y_k$ for $1 \le k \le n$ are in fact the classical "rays"

$$t \mapsto y_k \pm \frac{\xi_k}{|\xi_k|} (t-s) \qquad (1 \le k \le n).$$

In the general theory, each $a_{jh}^{(N)}(t, s, x, \xi)$ is taken independent of   y,   and is obtained by <u>integrating, along each bicharacteristic curve, an ordinary linear differential equation of the first order</u> (in the variable   t),   whose coefficients are determined when the $a_{jh}^{(\ell)}$ are known for $\ell \le N-1$; finally the induction starts with the values of the

$a_{jh}^{(o)}(s,s,x,\xi)$, which are given by the linear system

$$(66) \qquad \sum_{j=0}^{m-1} (iq'_j(s,x,\xi))^k a_{jh}^{(o)}(s,s,x,\xi) = \delta_{jh} \qquad (0 \leq k \leq m-1)$$

where $q'_j(s,x,\xi) = q_j(s,x,\text{grad}_x \psi_j(s,s,x,\xi))$; the determinant of that system is $\neq 0$ because the q_j have been supposed to be <u>distinct</u>.

One of the consequences of this remarkable construction is that, in the explicit formula (59) for $f = 0$, one may replace the "initial conditions" g_j by arbitrary <u>distributions</u> S_j on Ω; v is then replaced by a distribution T solution of $P \cdot T = 0$. For these equations, the "trace" T_t of such a distribution on the hyperplane $\{t\} \times \Omega$ may be defined (although T is not a function) as a distribution <u>on</u> Ω, varying with t, and which may be said to "propagate" with the "time" t, starting from the "initial" distribution $T_{t_o} =$ $= S_o$. It is then possible to show that the singular support of T_t is contained in a set M_t obtained in the following way: one considers the union M_o of the singular supports of the distributions S_j, and all the bicharacteristics issued from points of M_o; M_t is the set of all points on these bicharacteristics at time t. This gives a precise meaning to a phenomenon which had been well known for second order strictly hyperbolic equations, and particular types of "initial values": <u>the singularities propagate along the bi-characteristics</u>.

Under additional assumptions, it is possible to extend these results when in the decomposition (56), some of the q_j are equal [41].

II) Another important application of operators generalizing the pseudo-differential operators is the problem of <u>local e-xistence</u> of solutions for a partial differential equation $P \cdot u = f$, which has stimulated much research after H. Lewy's discovery (chap.II, §2). Over a period of more than 15 years, the combined efforts of Hörmander, Nirenberg and Trèves succeeded in formulating a system of conditions which were finally proved to be necessary and sufficient for local existence by Beals and Fefferman [20], using new types of operators [19]. We cannot here do more than refer the reader to these papers.

REFERENCES

[1] N.H. ABEL, Oeuvres, 2 vol. éd. Sylow et Lie, Christiania, 1881.

[2] N. ADASCH et al., Topological vector spaces, Lecture Notes in Math., nº 639, Berlin-Heidelberg-New York, Springer, 1978.

[3] N. AKHIEZER, The classical moment problem, Edinburg-London, Oliver and Boyd, 1965.

[4] E. AKIN, The metric theory of Banach manifolds, Lecture Notes in Math., nº 662, Berlin-Heidelberg-New York, Springer, 1978.

[5] L. ALAOGLU, Weak topologies of normed linear spaces, Ann. of Math., 41 (1940), p.252-267.

[5 bis] L. ALAOGLU, Weak convergence of linear functionals, Bull. Amer. Math. Soc., 44 (1938), p.196.

[6] P. ALEXANDROFF und P. URYSOHN, Zur Theorie der topologischen Räume, Math. Ann., 92 (1924), p.258-266.

[7] N. ARONSZAJN and K.T. SMITH, Invariant subspaces of completely continuous operators, Ann. of Math., 60 (1954), p.345-350.

[8] Giulio ASCOLI, Le curve limiti di una varietà data di curve, Mem. Acc. dei Lincei, (3), 18 (1883), p.521-586.

[9] Guido ASCOLI, Sugli spazi lineari metrici e le loro varietà lineari, Ann. di Mat., (4) 10 (1932), p.33-81 and 203-232.

[10] M. ATIYAH, Elliptic operators, discrete groups and von Neumann algebras, Astérisque, 32-33 (1976), p-43-72.

[11] R. BAIRE, Sur les fonctions de variables réelles, Ann. di Mat., (3), 3 (1899), p.1-123.

[12] S. BANACH, Sur les opérations dans les ensembles abstraits et leur application aux équations intégrales, Fund. Math., 3 (1923), p.133-181.

[13] S. BANACH, Sur le problème de la mesure, <u>Fund. Math.</u> 4
       (1923), p.7-33.

[14] S. BANACH, Sur les fonctionnelles linéaires, <u>Studia Math.</u>,
       1 (1929), p.211-216 et 223-229.

[15] S. BANACH, <u>Théorie des opérations linéaires</u>, Warszawa,
       1932.

[16] S. BANACH et H. STEINHAUS, Sur le principe de condensa-
       tion des singularités, <u>Fund. Math.</u>, 9 (1927),
       p.50-61.

[17] Banach spaces of analytic functions, Kent, Ohio, 1976
       Proceedings, <u>Lecture Notes in Math.</u>, nº 604, Berlin-
       Heidelberg-New York, Springer, 1977.

[18] K. BARBEY and H. KÖNIG, Abstract analytic function theory
       and Hardy algebras, <u>Lecture Notes in Math.</u>, nº 593,
       Berlin-Heidelberg-New York, Springer, 1977.

[19] R. BEALS, A general calculus of pseudodifferential opera-
       tors, <u>Duke Math. Journ.</u>, 42 (1975), p.1-42.

[20] R. BEALS and C. FEFFERMAN, Spatially inhomogeneous pseudo-
       differential operators I, <u>Comm. Pure Appl. Math.</u>,
       27 (1974), p.1-24.

[21] J. BENEDETTO, <u>Spectral synthesis</u>, New York, Academic
       Press, 1975.

[22] G.D. BIRKHOFF and O. KELLOGG, Invariant points in func-
       tion space, <u>Trans. Amer. Math. Soc.</u>, 23 (1922),
       p.96-115.

[23] M. BÔCHER, An introduction to the theory of integral
       equations, <u>Cambridge Tracts</u> nº 10, Cambridge Univ.
       Press, 1909.

[24] S. BOCHNER, Darstellung realvariabler und analytischer
       Funktionen durch verallgemeinerte Fourier- und
       Laplace-Integrale, <u>Math. Ann.</u>, 97 (1927), p.635-662.

[25] N. BOURBAKI, Sur les espaces de Banach, <u>C.R. Acad. Sci.</u>,
       206 (1938), p.1701-1704.

[26] N. BOURBAKI, Eléments de Mathématique, Livre V, Espaces
       vectoriels topologiques, <u>Actual. Scient. Ind.</u>,
       Chap.I-II, n⁰ 1189, Chap.III-V, n⁰ 1229, Hermann,
       Paris, 1953-55.

[27] N. BOURBAKI, Eléments de Mathématique, Théories spectrales,
       Chap.I-II, <u>Actual. Scient. Ind.</u>, n⁰ 1332, Hermann,
       Paris, 1967.

[28] N. BOURBAKI, <u>Eléments d'histoire des mathématiques</u>,
       Hermann, Paris, 1969.

[29] C. BOURLET, Sur les opérations en général, et les équa-
       tions différentielles d'ordre infini, <u>Ann. Ec. Norm.
       Sup.</u>, (3) 14 (1897), p.133-189.

[30] M. BRELOT, Historical introduction, <u>C.I.M.E. 1⁰ Ciclo
       1969, Potential Theory</u>, p.1-21, Roma, Cremonese,
       1970.

[31] M. BREUER, Fredholm theories in von Neumann algebras,
       <u>Math. Ann.</u>, 178 (1968) p.243-254, and 180 (1969),
       p.313-325.

[32] J.P. BREZIN, Harmonic Analysis on compact solvmanifolds,
       <u>Lecture Notes in Math.</u>, n⁰ 602, Berlin-Heidelberg-
       New York, Springer, 1977.

[33] A. BROWDER, <u>Introduction to function algebras</u>, New York-
       Amsterdam, Benjamin, 1969.

[34] F. BROWDER, The eigenfunction expansion theorem for the
       general self-adjoint singular elliptic partial dif-
       ferential operator, <u>Proc. Nat. Acad. Sci. USA</u>, 40
       (1954), p.454-459.

[35] H. BURKHARDT, Sur les fonctions de Green relatives à un
       domaine à une dimension, <u>Bull. Soc. Math. de France</u>,
       22 (1894), p.71-75.

[36] C*-algebras and applications to physics, Proceedings 1977,
       <u>Lecture Notes in Math.</u>, n⁰ 650, Berlin-Heidelberg-
       New York, Springer, 1978.

[37] T. CARLEMAN, Sur les équations intégrales singulières à noyau réel et symétrique, Uppsala, Univ. Arsskrift, 1923.

[38] T. CARLEMAN, Edition complète des articles, publiée par l'Institut Mittag-Leffler, Malmö, Litos Reprotryck, 1960.

[39] E. CARTAN, Leçons sur les invariants intégraux, Paris, Hermann, 1922.

[40] A.L. CAUCHY, Oeuvres complètes, 26 vol. (2 séries), Paris, Gauthier-Villars, 1882-1958.

[41] J. CHAZARAIN, Opérateurs hyperboliques à caractéristiques de multiplicité constante, Ann. Institut Fourier, 24, fasc. 1 (1974), p.173-202.

[42] G. CHOQUET, Unicité des représentations intégrales au moyen des points extrémaux dans les cônes convexes réticulés, C.R. Acad. Sci., 243 (1956), p.555-557; Existence des représentations intégrales dans les cônes convexes, Ibid., p.699-702 et 736-737.

[43] Conference on harmonic Analysis, College Park, Maryland, 1971, Lecture Notes in Math., n° 266, Berlin-Heidelberg-New York, Springer, 1972.

[44] A. CONNES, On the classification of von Neumann algebras and their automorphisms, Symposia math. 20 (1976), p.435-478.

[45] A. CONNES, Sur la théorie non commutative de l'intégration, in Lecture Notes in Math., n° 725, p.19-143, Berlin-Heidelberg-New York, Springer, 1979.

[46] P.J. DANIELL, Stieltjes-Volterra products, Congr. Intern. des Math., Strasbourg, 1920, p.130-136.

[47] J. DAY, Normed linear spaces, 3rd ed., Berlin-Heidelberg-New York, Springer, 1973.

[48] R. DEDEKIND, Gesammelte math. Werke, 3 vol., Braunschweig, Vieweg, 1932.

[49] G. DE RHAM, Über mehrfache Integrale, Abh. math. Sem.
     hansischen Univ., 12 (1938), p.313-339.

[50] J. DIESTEL, Geometry of Banach spaces, Lecture Notes in
     Math., n⁰ 485, Berlin-Heidelberg-New York, Springer,
     1975.

[51] J. DIEUDONNÉ, La dualité dans les espaces vectoriels to-
     pologiques, Ann. Ec. Norm. Sup., (3) 59 (1942),
     p.107-139.

[52] J. DIEUDONNÉ, Calcul infinitésimal, Paris, Hermann, 1968.

[53] J. DIEUDONNÉ and J. CARRELL, Invariant theory, old and
     new, New York-London, Academic Press, 1971.

[54] J. DIEUDONNÉ et al., Abrégé d'histoire des mathématiques,
     1700-1900, 2 vol. Paris, Hermann, 1978.

[55] J. DIEUDONNÉ et L. SCHWARTZ, La dualité dans les espaces
     (F) et (LF), Ann. Institut Fourier, 1 (1949),
     p.61-101.

[56] P.A.M. DIRAC, The physical interpretation of the quantum
     dynamics, Proc. Royal Soc. London, A, 113 (1926-1927),
     p.621-641.

[57] J. DIXMIER, Les algèbres d'opérateurs dans l'espace hil-
     bertien (algèbres de von Neumann), Paris, Gauthier-
     Villars, 1957.

[58] J. DIXMIER, Les C*-algèbres et leurs représentations,
     Paris, Gauthier-Villars, 1964.

[59] E. DUBINSKY, The structure of nuclear Fréchet spaces,
     Lecture Notes in Math., n⁰ 720, Berlin-Heidelberg-
     New York, Springer, 1979.

[60] P. DU BOIS-REYMOND, Erläuterungen zu den Anfangsgründen
     der Variationsrechnung, Math. Ann., 15 (1879),
     p.289-314 and 564-576.

[61] P. DU BOIS-REYMOND, Bemerkungen über $\Delta z = \dfrac{\partial^2 z}{\partial x^2} + \dfrac{\partial^2 z}{\partial y^2} = 0$,
     J. de Crelle, 103 (1888), p.204-229.

[62] N. DUNFORD and J. SCHWARTZ, Linear operators, 3 vol.,
       New York-London-Sydney-Toronto, Wiley-Interscience,
       1958-1971.

[63] L. EHRENPREIS, Solutions of some problems of division I,
       Amer. Journ. of Math., 76 (1954), p.883-903.

[64] L. ERDELYI and R. LANGE, Spectral decompositions on
       Banach spaces, Lecture Notes in Math., n⁰ 623,
       Berlin-Heidelberg-New York, Springer, 1977.

[65] L. EULER, Opera Omnia, 61 vol. parus (4 séries), Leipzig-
       Berlin-Zürich, Teubner et O. Füssli, 1911-1980.

[66] E. FISCHER, Sur la convergence en moyenne, C.R. Acad.
       Sci., 144 (1907), p.1022-1024; Application d'un
       théorème sur la convergence en moyenne, ibid.,
       p.1148-1151.

[67] J.B. FOURIER, Oeuvres, 2 vol., Paris, Gauthier-Villars,
       1888-1890.

[68] Fourier integral operators and partial differential
       equations, Colloque international, Université de
       Nice, 1974, Lecture Notes in Math., n⁰ 459, Berlin-
       Heidelberg-New York, Springer, 1975.

[69] M. FRÉCHET, Généralisation d'un théorème de Weierstrass,
       C.R. Acad. Sci., 139 (1904), p.848-850.

[70] M. FRÉCHET, Sur les opérations linéaires, Trans. Amer.
       Math. Soc., 6 (1905), p.134-140.

[71] M. FRÉCHET, Sur quelques points du Calcul fonctionnel,
       Rend. Circ. mat. Palermo, 22 (1906), p.1-74.

[72] M. FRÉCHET, Essai de géométrie analytique à une infinité
       de coordonnées, Nouv. Ann. de Math., (4) 8 (1908),
       p.97-116 and 289-317.

[73] M. FRÉCHET, Les espaces abstraits topologiquement affi-
       nes, Acta math., 47 (1926), p.25-52.

[74] I. FREDHOLM, Oeuvres complètes publiées par l'Institut
       Mittag-Leffler, Malmö, Litos Reprotryck, 1955.

[75] I. FREDHOLM, Sur une classe d'équations fonctionnelles,
         Acta math., 27 (1903), p.365-390.

[76] K. FRIEDRICHS, On differential operators in Hilbert
         spaces, Amer. Journ. of Math., 61 (1939), p.523-544.

[77] K. FRIEDRICHS, On the differentiability of the solutions
         of linear elliptic differential equations, Comm.
         Pure Appl. Math., 6 (1953), p.299-325.

[78] G. FROBENIUS, Gesammelte Abhandlungen, 3 vol., Berlin-
         Heidelberg-New York, Springer, 1968.

[79] T. GAMELIN, Uniform algebras, Englewood Cliffs, N.J.,
         Prentice-Hall, 1969.

[80] L. GÅRDING, Dirichlet's problem for linear elliptic par-
         tial differential equations, Math. Scand., 1 (1953),
         p.55-72.

[81] L. GÅRDING, Eigenfunction expansions connected with
         elliptic differential operators, C.R. du 12e Congrès
         des mathém. scand., 1953, Lund, H. Ohlssons
         Boktryckeri, 1954.

[82] C.F. GAUSS, Werke, 12 vol. Göttingen, 1870-1927.

[83] I. GELFAND, Normierte Ringen, Mat. Sborn., (N.S.) 9
         (1941), p.3-24.

[84] I. GELFAND and B. LEVITAN, On the determination of a dif-
         ferential equation from its spectral function, Amer.
         Math. Soc. Transl., (2) 1 (1955) p.253-304.

[85] I. GELFAND and M. NAIMARK, On the imbedding of normed
         rings into the ring of operators in Hilbert space,
         Mat. Sborn. (N.S.), 12 (1943), p.197-213.

[86] I. GELFAND and D. RAIKOV, On the theory of characters of
         commutative topological groups, Dokl. Akad. Nauk,
         28 (1940), p.195-198.

[87] I. GELFAND and G. SHILOV, Generalized functions I, New
         York-London, Academic Press, 1964.

[88] H. GOLDSTINE, Weakly complete Banach spaces, <u>Duke math.</u>
      <u>Journ.</u>, 4 (1938), p.126-131.

[89] J.P. GRAM, Ueber die Entwickelung reeller Functionen in
      Reihen mittelst der Methode der kleinsten Quadrate,
      <u>J. de Crelle</u>, 94 (1883), p.41-73.

[90] G. GREEN, <u>Mathematical Papers</u>, Paris, Hermann, 1903.

[91] A. GROTHENDIECK, Produits tensoriels topologiques et es-
      paces nucléaires, <u>Mem. Amer. Math. Soc.</u>, nº 16,
      Providence, Amer. Math. Soc., 1953.

[92] A. GROTHENDIECK, <u>Espaces vectoriels topologiques</u>, 2e éd.,
      São Paulo, éd. Soc. mat. de São Paulo, 1958.

[93] J. HADAMARD, <u>Le problème de Cauchy et les équations aux</u>
      <u>dérivées partielles linéaires hyperboliques</u>, Paris,
      Hermann, 1932.

[94] J. HADAMARD, <u>Oeuvres</u>, 4 vol., Paris, Ed. du C.N.R.S.,
      1968.

[95] H. HAHN, Über die Integrale des Herrn Hellinger und die
      orthogonalinvarianten der quadratischen Formen von
      unendlichvielen Veränderlichen, <u>Monatsh. für Math.</u>
      <u>und Phys.</u>, 23 (1912), p.161-224.

[96] H. HAHN, Über eine Verallgemeinerung der Fourierschen
      Integralformel, <u>Acta math.</u>, 49 (1926), p.301-353.

[97] H. HAHN, Über Folgen linearer Operationen, <u>Monatsh. für</u>
      <u>Math. und Phys.</u>, 32 (1922), p.1-88.

[98] H. HAHN, Über lineare Gleichungssysteme in linearen
      Räumen, <u>J. de Crelle</u>, 157 (1927), p.214-229.

[99] H. HAMBURGER, Über die Zerlegung des Hilbertschen Raumes
      durch vollstetige lineare Transformationen, <u>Math.</u>
      <u>Nachr.</u>, 4 (1950), p.56-69.

[100] F. HAUSDORFF, <u>Grundzüge der Mengenlehre</u>, Leipzig, Veit,
      1914.

[101] F. HAUSDORFF, <u>Mengenlehre</u>, Berlin, de Gruyter, 1927.

288   REFERENCES

[102] F. HAUSDORFF, Zur Theorie der linearen metrischen Räume,
       J. de Crelle, 167 (1931), p.294-311.

[103] T. HAWKINS, Lebesgue's theory of integration, Madison-
       Milwaukee-London, The Univ. of Wisconsin Press, 1970.

[104] E. HEINE, Handbuch der Kugelfunctionen, 2 vol., Berlin,
       Reimer, 1881.

[105] E. HELLINGER, Neue Begründung der Theorie quadratischer
       Formen von unendlichvielen Veränderlichen, J. de
       Crelle, 136 (1909), p.210-271.

[106] E. HELLINGER und O. TOEPLITZ, Grundlagen für eine
       Theorie der unendlichen Matrizen, Göttinger Nachr.,
       1906, p.351-355 and Math. Ann., 69 (1910), p.281-330.

[107] E. HELLINGER und O. TOEPLITZ, Integralgleichungen und
       Gleichungen mit unendlichvielen Unbekannten, New
       York, Chelsea, 1953 (= Enzykl. der math. Wiss.,
       II C 13, 1927).

[107 bis] E. HELLY, Über lineare Funktionaloperationen,
       Sitzungsber. der math. naturwiss. Klasse der Akad.
       der Wiss. (Wien), 121 (1912), p.265-297.

[108] E. HELLY, Ueber Systeme linearer Gleichungen mit un-
       endlich vielen Unbekannten, Monatsh. für Math. und
       Phys., 31 (1921), p.60-91.

[109] G. HERGLOTZ, Über die Integration linearer, partieller
       Differentialgleichungen mit konstanten Koeffizienten,
       Abh. math. Sem. hansischen Univ., 6 (1928), p.189-
       197.

[110] E.HILB, Über Integraldarstellungen willkürlicher Funk-
       tionen, Math. Ann., 66 (1908), p.1-66.

[111] D. HILBERT, Gesammelte Abhandlungen, 3 vol. Berlin,
       Springer, 1932-1935.

[112] D. HILBERT, Grundzüge einer allgemeinen Theorie der
       Integralgleichungen, 2^eéd., Leipzig-Berlin, Teubner,
       1924.

[113] Hilbert space operators, Proceedings 1977, <u>Lecture No-</u>
       <u>tes in Math</u>., n° 693, Berlin-Heidelberg-New York,
       Springer, 1978.

[114] G.W. HILL, <u>Collected mathematical works</u>, 4 vol.,
       Carnegie Inst., Washington, 1905-1907.

[115] E. HILLE and R. PHILLIPS, <u>Functional Analysis and semi-</u>
       <u>groups</u>, Amer. Math. Soc. Coll. Publ. XXXI, 1957.

[116] K. HOFFMAN, <u>Banach spaces of analytic functions</u>, Engle-
       wood Cliffs (N.J.), Prentice Hall, 1962.

[117] L. HÖRMANDER, <u>Linear partial differential operators</u>,
       Berlin-Heidelberg-New York, Springer, 1964.

[118] L. HÖRMANDER, On the division of distributions by poly-
       nomials, <u>Ark. Mat</u>. 3 (1958), p. 555-568.

[119] L. HÖRMANDER, On the existence and the regularity of
       solutions of linear pseudo-differential equations,
       <u>L'Enseignement math</u>., (2), 17 (1971), p.99-163.

[120] C.G.J. JACOBI, <u>Gesammelte Werke</u>, 7 vol., Berlin, Reimer,
       1881-1891.

[121] M. JAMMER, <u>The conceptual development of quantum me-</u>
       <u>chanics</u>, New York, McGraw Hill, 1966.

[122] F. JOHN, <u>Plane waves and spherical means applied to</u>
       <u>partial differential equations</u>, New York-London,
       Interscience, 1955.

[123] C. JORDAN, <u>Traité des substitutions et des équations</u>
       <u>algébriques</u>, 2e éd., Paris, Gauthier-Villars et
       A. Blanchard, 1957.

[124] T. KATO, <u>Perturbation theory for linear operators</u>,
       Berlin-Heidelberg-New York, Springer, 1966.

[125] J. KELLEY <u>et al</u>., <u>Linear topological spaces</u>, Princeton-
       Toronto-New York-London, Van Nostrand, 1963.

[126] K. KODAIRA, On ordinary differential equations of any
       even order and the corresponding eigenfunction
       expansions, <u>Amer. Journ. of Math</u>., 72 (1950),
       p.502-544.

[127] G. KÖTHE, Neubegründung der Theorie der vollkommenen
        Räume, Math. Nachr., 4 (1950), p.70-80.

[128] G. KÖTHE, Topological vector spaces, 2 vol., Berlin-
        Heidelberg-New York, Springer, 1969-1979.

[129] G. KÖTHE und O. TOEPLITZ, Lineare Räume mit unendlich-
        vielen Koordinaten und Ringe unendlicher Matrizen,
        J. de Crelle, 171 (1934), p.193-226.

[130] A. KOLMOGOROFF, Zur Normierbarkeit eines allgemeinen
        topologischen linearen Raumes, Studia Math., 5 (1934),
        p.29-33.

[131] M. KREIN and D. MILMAN, On extreme points of regular
        convex sets, Studia Math., 9 (1940), p.133-138.

[132] K-Theory and operator algebras, Athens, Georgia 1975,
        Lecture Notes in Math., nº 575, Berlin-Heidelberg-
        New York, Springer, 1977.

[133] J.PH. LABROUSSE, Thèse, Univ. de Nice, 1979.

[134] H. LACEY, The isometric theory of classical Banach
        spaces, Berlin-Heidelberg-New York, Springer, 1974.

[135] J.L. LAGRANGE, Oeuvres, 14 vol., Paris, Gauthier-Villars,
        1867-1892.

[136] E. LANDAU, Über einen Konvergenzsatz, Göttinger Nachr.,
        1907, p.25-27.

[137] P.S. LAPLACE, Oeuvres, 14 vol., Paris, Gauthier-Villars,
        1878-1912.

[138] H. LEBESGUE, Oeuvres scientifiques, 5 vol., Genève,
        L'enseignement math., 1972-1973.

[139] H. LEBESGUE, Leçons sur les séries trigonométriques,
        Paris, Gauthier-Villars, 1906.

[140] G. LEIBOWITZ, Lectures on complex function algebras,
        Glenview (Ill.), Scott Foresman, 1970.

[141] J. LERAY, Sur le mouvement d'un fluide visqueux
        emplissant l'espace, Acta math., 63 (1934),
        p.193-248.

[142] J. LERAY et J. SCHAUDER, Topologie et équations
        fonctionnelles, Ann. Ec. Norm. Sup., (3), 51 (1934),
        p.43-78.

[143] J. LE ROUX, Sur les intégrales des équations linéaires
        aux dérivées partielles du 2e ordre à 2 variables
        indépendantes, Ann. Ec. Norm. Sup., (3) 12 (1895),
        p.227-316.

[144] E. LE ROY, Sur l'intégration de l'équation de la chaleur
        (2e partie), Ann. Ec. Norm. Sup., (3) 15 (1898),
        p.9-178.

[145] E.E. LEVI, Opere, 2 vol., Roma, Cremonese, 1959-1960.

[146] P. LÉVY, Leçons d'Analyse fonctionnelle, Paris, Gauthier-
        Villars, 1922.

[147] A. LIAPOUNOV, Sur certaines équations qui se rattachent
        au problème de Dirichlet, Journ. de Math., (5) 4
        (1898), p.241-311.

[148] A. LIAPOUNOV, Sur une proposition de la théorie des
        probabilités, Izv. Akad. Nauk, (5) 13 (1900),
        p.359-386.

[149] J. LINDENSTRAUSS and L. TZAFRIRI, Classical Banach
        spaces, Lecture Notes in Math., nº 338, Berlin-
        Heidelberg-New York, Springer, 1973.

[150] J. LINDENSTRAUSS and L. TZAFRIRI, Classical Banach spaces
        I, Berlin-Heidelberg-New York, Springer, 1977.

[151] J. LIOUVILLE, Sur le développement des fonctions ou
        parties de fonctions en séries dont les divers
        termes sont assujettis à satisfaire à une même
        équation différentielle du second ordre contenant
        an paramètre arbitraire, Journ. de Math., (1) 1
        (1836) p.253-265 and 2 (1837), p.16-35 and 418-436.

[152] R. LIPSMAN, Group representations. A survey of some
        current topics, Lecture Notes in Math., nº 388,
        Berlin-Heidelberg-New York, Springer, 1974.

[153] S. ŁOJASIEWICZ, Sur le problème de division, <u>Studia
        Math.</u>, 18 (1959), p.87-136.

[154] G. MACKEY, On convex topological spaces, <u>Trans. Amer.
        Math. Soc.</u>, 60 (1946), p.519-537.

[155] G. MACKEY, Harmonic Analysis as the exploitation of
        symmetry - a historical survey, in History of
        Analysis, <u>Rice Univ. Studies</u>, 64 (1978) n°s 2 et 3
        p.73-228.

[156] B. MALGRANGE, Existence et approximation des solutions
        des équations aux dérivées partielles et des équa-
        tions de convolution, <u>Ann. Institut Fourier</u>, 6
        (1955-1956), p.271-355.

[157] F. MAUTNER, On eigenfunction expansions, <u>Proc. Nat. Ac.
        Sci. USA</u>, 39 (1952), p.49-53.

[158] S. MAZUR, Über konvexe Mengen in linearen normierten
        Räumen, <u>Studia Math.</u>, 4 (1933), p.70-84.

[159] S. MAZUR und W. ORLICZ, Über Folgen linearer Operationen,
        <u>Studia Math.</u>, 4 (1933), p.152-157.

[160] S. MAZUR et W. ORLICZ, Sur les espaces métriques liné-
        aires (I), <u>Studia Math.</u>, 10 (1948), p.184-208.

[161] H. MINKOWSKI, <u>Gesammelte Abhandlungen</u>, 2 vol., Leipzig-
        Berlin, Teubner, 1911.

[162] H. MINKOWSKI, <u>Geometrie der Zahlen</u>, Leipzig, Teubner,
        1896.

[163] E.H. MOORE, Introduction to a form of general Analysis,
        <u>The New Haven Math. Colloquium</u>, New Haven, Yale
        Univ. Press, 1910.

[164] E.H. MOORE and H.L. SMITH, A general theory of limits,
        <u>Amer. Journ. of Math.</u>, 44 (1922), p.102-121.

[165] C. NEUMANN, <u>Untersuchungen über das logarithmische und
        Newton'sche Potential</u>, Leipzig, Teubner, 1877.

[166] M. NEUMARK, <u>Lineare Differentialoperatoren</u>, Berlin,
        Akad. Verlag, 1960.

[167] L. NIRENBERG, Remarks on strongly elliptic partial dif-
      ferential equations, Comm. Pure Appl. Math., 8 (1955),
      p.648-674.

[168] L. NIRENBERG, Estimates and existence of solutions for
      elliptic equations, Comm. Pure Appl. Math., 9
      (1956), p.509-529.

[169] L. NIRENBERG, Lectures on linear partial differential
      equations, CBMS Reg. Conf. Series in Math., 17
      (1973), Providence, Amer. Math. Soc.

[170] W. OSGOOD, Non uniform convergence and the integration
      of series term by term, Amer. Journ. of Math., 19
      (1897), p.155-190.

[171] J. PEETRE, Another approach to elliptic boundary value
      problems, Comm. Pure Appl. Math., 14 (1961),
      p.711-731.

[172] E. PICARD, Oeuvres, vol. II, Paris, Ed. du C.N.R.S.,
      1979.

[173] S. PINCHERLE, Opere scelte, 2 vol., Roma, Cremonese,
      1954.

[174] M. PLANCHEREL, Contributions à l'étude de la représen-
      tation d'une fonction arbitraire par des intégrales
      définies, Rend. Circ. mat. Palermo, 30 (1910),
      p.289-335.

[175] J. PLEMELJ, Zur Theorie der Fredholmsche Funktional-
      gleichung, Monatsh. für Math. und Phys., 15 (1904),
      p.93-128.

[176] A. PLIŠ, A smooth linear elliptic equation without any
      solution in a sphere, Comm. Pure Appl. Math., 14
      (1961), p.599-616.

[177] H. POINCARÉ, Oeuvres, 11 vol., Paris, Gauthier-Villars,
      1916-1956.

[178] D. POISSON, Remarques sur une équation qui se présente
      dans la théorie de l'attraction des sphéroïdes,
      Bull. Soc. Philomath. Paris, 3 (1813), p.388-392.

[179] L. PONTRJAGIN, The theory of topological commutative
       groups, Ann. of Math., 35 (1934), p.361-388.

[180] H. PRÜFER, Neue Herleitung der Sturm-Liouvillesche
       Reihenentwicklung stetigen Funktionen, Math. Ann.,
       95 (1926), p.499-518.

[181] F. PRYM, Zur Integration der Differentialgleichung
       $\frac{\partial^2 u}{\partial x^2} + \frac{\partial^2 u}{\partial y^2} = 0$, J. de Crelle, 73 (1871), p.340-364.

[182] B. RIEMANN, Gesammelte mathematische Werke, 2e éd.,
       Leipzig, Teubner, 1892; Nachträge, ibid., 1902.

[183] F. RIESZ, Oeuvres complètes, 2 vol., Paris, Gauthier-
       Villars, 1960.

[184] F. RIESZ, Les systèmes d'équations linéaires à une in-
       finité d'inconnues, Paris, Gauthier-Villars, 1913.

[185] J.J. SCHÄFFER, Geometry of spheres in normed spaces,
       New York, Dekker, 1976.

[186] J. SCHAUDER, Zur Theorie stetiger Abbildungen in
       Funktionalräumen, Math. Zeitschr., 26 (1927),
       p.47-65 and 417-431.

[187] J. SCHAUDER, Der Fixpunktsatz in Funktionalräumen,
       Studia Math., 2 (1930), p.171-180.

[188] J. SCHAUDER, Über lineare, vollstetige Operationen,
       Studia Math., 2 (1930), p.183-196.

[189] J. SCHAUDER, Über den Zusammenhang zwischen der
       Eindeutigkeit und Lösbarkeit partieller Differen-
       tialgleichungen zweiter Ordnung vom elliptischen
       Typus, Math. Ann., 106 (1932), p.661-721.

[190] J. SCHAUDER, Das Anfangswertproblem einer quasilinearen
       hyperbolischen Differentialgleichung zweiter Ordnung
       in beliebiger Anzahl von unabhängigen Veränderlichen,
       Fund. Math., 24 (1935), p.213-246.

[191] E. SCHMIDT, Zur Theorie der linearen und nichtlinearen
        Integralgleichungen. I. Teil: Entwickelung will-
        kürlicher Funktionen nach Systeme nachgeschriebener,
        Math. Ann., 63 (1907), p.433-476.

[192] E. SCHMIDT, Ueber die Auflösung linearer Gleichungen mit
        unendlich vielen Unbekannten, Rend. Circ. mat.
        Palermo, 25 (1908), p.53-77.

[193] I. SCHUR, Gesammelte Abhandlungen, 3 vol., Berlin-
        Heidelberg-New York, Springer, 1973.

[194] L. SCHWARTZ, Théorie des distributions, Actual. Scient.
        Ind., n°s 1091 and 1122, Paris, Hermann, 1950-1951.

[195] L. SCHWARTZ, Théorie des noyaux, Proc. Intern. Congress
        of mathem., Cambridge, Mass., 1950, vol.I, p.220-230,
        Providence, Amer. Math. Soc., 1952.

[196] H.A. SCHWARZ, Gesammelte mathematische Abhandlungen,
        2 vol., Berlin, Springer, 1890.

[197] R. SEELEY, Elliptic singular integral equations, Proc.
        Symp. Pure Math. X, 1967, p.308-313.

[198] P. SHANAHAN, The Atiyah-Singer index theorem, Lecture
        Notes in Math., n° 638, Berlin-Heidelberg-New York,
        Springer, 1978.

[199] I.M. SINGER, Future extensions of index theory and
        elliptic operators, Prospects in mathematics, Ann.
        Math. Studies n° 70, p.171-185, Princeton Univ.Press,
        1971.

[200] S. SOBOLEV, Méthode nouvelle à résoudre le problème de
        Cauchy pour les équations linéaires hyperboliques
        normales, Mat. Sborn. (N.S.), 1 (1936), p.39-72.

[201] S. SOBOLEV, Sur un théorème d'Analyse fonctionnelle
        (Russian), Mat. Sborn., (N.S.) 4 (1938), p.471-496.

[202] H. STEINHAUS, Additive und stetige Funktionaloperationen,
        Math. Zeitschr., 5 (1919), p.186-221.

[203] W. STEKLOFF, Sur les problèmes fondamentaux de la
       Physique mathématique, Ann. Ec. Norm. Sup., (3) 19
       (1902), p.192-259 and 455-490.

[204] W. STEKLOFF, Théorie générale des fonctions fondamenta-
       les, Ann. Fac. Sci. de Toulouse, (2) 6 (1904),
       p.351-475.

[205] T. STIELTJES, Recherches sur les fractions continues,
       Ann. Fac. Sci. de Toulouse, 8 (1894), p.J1-J122.

[206] M.H. STONE, Linear transformations in Hilbert space:
       I. Geometrical aspects, Proc. Nat. Acad. Sci. USA,
       15 (1929), p.198-200; II. Analytical aspects,
       ibid., p.423-425; III. Operational methods and
       group theory, ibid., 16 (1930), p.172-175.

[207] M.H. STONE, Linear transformations in Hilbert space,
       Amer. Math. Soc. Coll. Publ. XV, 1932.

[208] M.H. STONE, The theory of representation for Boolean
       algebras, Trans. Amer. math. Soc., 40 (1936),
       p.37-111.

[209] C. STURM, Sur les équations différentielles linéaires
       du second ordre, Journ. de Math., (1) 1 (1836),
       p.106-186.

[210] M. TAKESAKI, Tomita's theory of modular Hilbert alge-
       bras and its applications, Lecture Notes in Math.,
       nº 128, Berlin-Heidelberg-New York, Springer, 1970.

[211] P. TCHEBYCHEF, Oeuvres, 2 vol., St-Petersbourg,
       1899-1907.

[212] E. TITCHMARSH, Eigenfunction expansions associated with
       second-order differential equations, 2 vol. Oxford,
       Clarendon Press, 1946-1958.

[213] O. TOEPLITZ, Die Jacobische Transformation der quadra-
       tischen Formen von unendlichvielen Veränderlichen,
       Göttinger Nachr., 1907, p.101-109.

[214] O. TOEPLITZ, Zur Theorie der quadratischen Formen von
        unendlichvielen Veränderlichen, Göttinger Nachr.,
        1910, p.489-506.

[215] F. TRÈVES, Topological vector spaces, distributions and
        kernels, New York, Academic Press, 1967.

[216] E. VAN KAMPEN, Locally bicompact abelian groups and
        their character groups, Ann. of Math., 36 (1935),
        p.448-463.

[217] V.S. VARADARAJAN, Harmonic Analysis on real reductive
        groups, Lecture Notes in Math., nº 576, Berlin-
        Heidelberg-New York, Springer, 1977.

[218] M. VIŠIK, On general boundary problems for elliptic
        differential equations (Russian), Trudy Moskov. Mat.
        Obsc., 1 (1952), p.187-246.

[219] V. VOLTERRA, Opere matematiche, 5 vol., Acc. dei Lincei,
        1954-1962.

[220] H. VON KOCH, Sur la convergence des déterminants d'ordre
        infini, Bihang till Vet. Akad. Handlinger, Afd. 1,
        nº 4, 1896.

[221] J. VON NEUMANN, Collected Works, 6 vol., Oxford-London-
        New York-Paris, Pergamon Press, 1961-1963.

[222] L. WAELBROECK, Le calcul symbolique dans les algèbres
        commutatives, Journ. de Math., (9) 33 (1954),
        p.147-186.

[223] G. WARNER, Harmonic Analysis on semi-simple groups,
        2 vol., Berlin-Heidelberg-New York, Springer, 1972.

[224] H. WEBER, Über die Integration der partiellen Differ-
        entialgleichung $\frac{\partial^2 u}{\partial x^2} + \frac{\partial^2 u}{\partial y^2} + k^2 u = 0$. Math. Ann., 1
        (1869), p.1-36.

[225] H. WEBER, Lehrbuch der Algebra, 2e éd., 3 vol.,
        Braunschweig, Vieweg, 1898-1908.

[226] A. WEIL, L'intégration dans les groupes topologiques et
        ses applications, Actual. Scient. et Ind., nº 869,
        Paris, Hermann, 1940.

[227] H. WEYL, <u>Gesammelte Abhandlungen</u>, 4 vol., Berlin-
       Heidelberg-New York, Springer, 1968.

[228] E. WEYR, Zur Theorie der bilinearen Formen, <u>Monatsh. für
       Math. und Phys.</u>, 1 (1890), p.163-236.

[229] N. WIENER, On the representation of functions by trigo-
       nometric integrals, <u>Math. Zeitschr.</u>, 24 (1926),
       p.575-616.

[230] W. WIRTINGER, Beiträge zur Riemann's Integrationsmethode
       für hyperbolische Differentialgleichungen, und deren
       Anwendungen auf Schwingungsprobleme, <u>Math. Ann.</u>, 48
       (1897), p.364-389.

[231] L.C. YOUNG, Generalized curves and the existence of an
       attained absolute minimum in the calculus of
       variations, <u>C.R. Soc. Sci. Varsovie</u>, 30 (1937),
       p.212-234.

[232] S. ZAREMBA, Sur l'équation aux dérivées partielles
       $\Delta u + \lambda u + f = 0$ et sur les fonctions harmoniques,
       <u>Ann. Ec. Norm. Sup.</u>, (3) 16 (1899), p.427-464.

[233] S. ZAREMBA, Sur un problème toujours possible comprenant,
       à titre de cas particulier, le problème de Dirichlet
       et celui de Neumann, <u>Journ. de Math.</u>, (9), 6 (1927),
       p.127-163.

[D]   J. DIEUDONNÉ, <u>Eléments d'Analyse</u>, vol. VII et VIII,
       Paris, Gauthier-Villars, 1978.

[S]   <u>A Source book in Classical Analysis</u> (ed. G. Birkhoff),
       Cambridge, Mass., Harvard Univ. Press, 1973.

AUTHOR INDEX

SUBJECT INDEX

Abel's integral equation: 92

Action of a group: 196

Adjoint of a linear differential operator: 10

Adjoint of an operator: 86,221

Adjoint of an unbounded operator: 174

Alexander's horned sphere: 69

*-algebra: 182

Alternating process of Schwarz: 40

A priori inequalities: 59,236

Asymptotic expansion of a pseudo-differential operator: 262

Axes of a quadratic form: 90,107

Baire's theorem: 141

Banach algebra: 185

Banach space: 143

Beer-Neumann integral equation: 42

Bessel's identity, Bessel's inequality: 21,108,109,111

Bicharacteristic curves: 277

Boundary conditions: 245

Bounded bilinear form: 113

Bounded set: 216,217

Buniakowsky's inequality: 51

Calderon operator: 270

C*-algebra: 187

Carleman kernel: 168,239

Carleman operator: 168,238

Cauchy-Kowalewska theorem: 26

Cayley transform: 180

Character of a Banach algebra: 186

Character of a group: 196,203

Characteristic hypersurface: 275

Closable operator: 174

Closed graph theorem: 142

Commutant of a set: 182

Printed and bound by CPI Group (UK) Ltd, Croydon, CR0 4YY

03/10/2024

01040333-0013